DETERMINED TO BE EXTRAORDINARY

Spectacular Stories of Modern Women in STEM

EDITED BY: DAWN HEIMER

Paperback ISBN: 979-8-218-27542-6

Hardcover ISBN: 979-8-218-27541-9

Library of Congress Control Number: 2024902298

Cover art by Eduardo Urbano Merino at www.eduardourbanomerino.com

For more information and resources and to engage with our community, please visit dawnheimer.com. Your journey in STEM empowerment continues beyond these pages.

DEDICATION

IN LOVING MEMORY OF LAURIE HALLORAN

Your unwavering dedication to advancing clinical research and your passion for leadership development and making a difference in the lives of patients have left an indelible mark on all who had the privilege of knowing you. Your courage, determination, and unwavering commitment to excellence serve as a guiding light for all aspiring women in STEM. May your spirit continue to guide us as we honor your memory and carry forward your legacy with pride and gratitude, and may your life's work inspire generations to come.

TABLE OF CONTENTS

01
CHRISTINA GOETHEL

07
LOLA ADEYEMI

10
ERIKA EBBEL ANGLE

Erika has been featured on Lifetime TV and Nova Science Now on "The Secret Lives of Scientists and Engineers."

17
EVA SUARTHANA

"I will continue teaching and doing research, as education is the key to strengthening our communities and the younger generations."

22
ANNE CAMILLE TALLEY

Volunteering helped Anne broaden and deepen her understanding of teams, mentors and advocates.

25
MELANIE L. FLORES

29
TRIKA GERARD

TABLE OF CONTENTS

33
SANDY JO MACARTHUR

Sandy Jo created a school within the LAPD police academy that serves underprivileged communities and prepares students for college and public service careers.

37
FELECIA BOWSER

Felecia is only one of two women and one of two minorities in her region to hold the post of Meteorologist-in-Charge.

39
KIMBERLY SMITH

44
SELINA WEBB

Give back when you can, and you will be rewarded beyond measure.

46
LAURIE HALLORAN

"Trying and failing at something new is much better than sticking with something that does not make you happy."

54
NIKITA ANGANE

"The world needs women like us: Leaders who do not doubt themselves."

57
LAURA GUERTIN

Laura was one of 100 female scientists selected for the INSIGHT into Diversity Inspiring Women in STEM Award.

61
MICHELE FINN

Positive and negative input made her stronger, better and more resilient. As a woman in a traditionally male-dominated positions, she needed to meet or exceed the standards set for her male peers.

65
SANDRA LOPEZ LEON

Sandra was recognized in the US as "101 most influential Latinos" in 2022.

68
ERIN SAWYER

TABLE OF CONTENTS

70
ROKELLE SUN

74
EMMA TOWLSON

Most of the outstanding mentors in Emma's professional life were people she approached out of admiration for their work, and she set out to develop working relationships with them.

79
DARCY GAGNE

You will stand apart from the crowd if you are willing to do what no one else will. Always take things to the next level, and don't let anyone tell you can't - including yourself.

83
LATHA PARVATANENI

87
ELSA SALAZAR CADE

Elsa has been named one of the ten best science teachers in the United States by the National Science Teachers Association.

97
CANDICE HUGHES

Candice has been the CEO and Founder of multiple venture companies.

102
ADRIANA L. ROMERO-OLIVARES

106
CHARLOTTE SIBLEY

Follow the ABCs: Aim for progress, not perfection; Be curious; have the Courage to speak up and let people hear your ideas.

TABLE OF CONTENTS

111

KARIN HOLLERBACH

"Keep checking in with yourself, and if you're not doing what you want, revise your goals and come up with a plan on how to get there."

114

FATEMEH RAZJOUYAN

LOOK NO FURTHER

The female STEM role models you've been looking for are right here, right now, to inspire you!

So many STEM books for young girls are history books. This book is full of living, real-life examples of women who are succeeding in STEM. Inspiring women leaders in STEM have come together to contribute to this collective memoir. They have original, passionate stories that describe their perseverance, spirit, brilliance, success, commitment, personal transformation, and growth through words and photographs.

They have chosen to contribute for some of the following reasons: they believe their life experiences are inspirational to young girls, sharing their story advances ideas they care about, they want to highlight the vital work they have accomplished, and they want to elevate their field of study.

This book will show you that you can pursue ambitious goals and succeed just like us! My 10-year-old and I were frustrated with the quality of STEM books. Most of the women in those books were dead (not contemporary), the stories were written by someone else, and all the books were illustrated. We felt strongly that this downplayed their work's significance (by portraying them always as cartoons). Also, as a woman in STEM and an abstract photographer, I sought a way to highlight the fantastic STEM women I work with. We are often invisible. Yet, you bump into us every day but don't realize the extraordinary scientific work we do because we look just like everyone else and don't always have a forum to tell our stories. I felt their stories needed to be told in their own words, with pictures, and I wanted to give them a voice.

When I was young and trying to decide my path in life, there were very few STEM female role models to look up to. I hope this book changes that for you!

For more information and resources and to connect, visit my website at **dawnheimer.com**

DAWN HEIMER
EDITOR

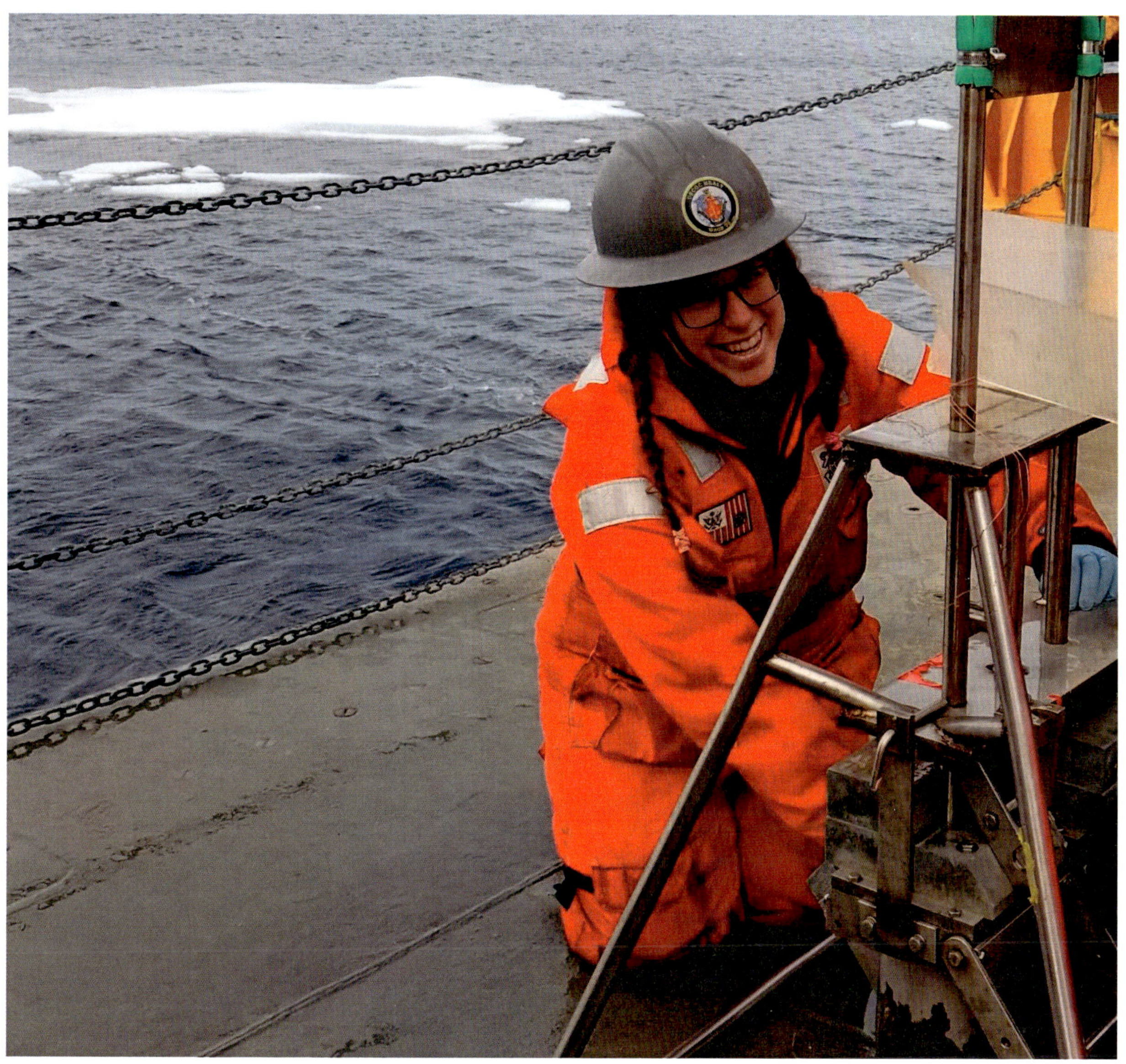

ASSISTANT PROFESSOR
ASSISTANT RESEARCH SCIENTIST

CHRISTINA GOETHEL

"The people you know and the places you see shape you into the person you dare to become. With a mix of high aspirations, persistence in the face of obstacles, and belief in yourself, you can become that daring person."

The first two ingredients to becoming the person you want to be are persistence and belief in yourself. When I was little, I used to walk up and down the beach in New Hampshire with my aunt. She taught me about the ocean and everything that lived in it. I loved those days. I had many questions about the animals, plants, and rocks. I wanted to learn more about the ocean and understand how it worked, but I did not realize how much until a wonderful opportunity came along when I was fourteen years old.

I was invited to take a trip to the Arctic as part of the People-to-People Student Ambassadors. I had already participated in two other trips with them. This trip would highlight one of my greatest passions, the ocean and the animals that called it home. I never thought I would go from the beaches of New Hampshire to a ship in the Arctic, but that is what happened. It was the trip of a lifetime! We would sail around Iceland, Greenland, and northern Canada for two weeks. We would be with experts in whales, glaciers, chemistry, and tiny plants called phytoplankton. I wanted to go and learn everything I could, but with two disabled parents and little money, I didn't know how I could pay for another trip. Also, my mom is terrified of the ocean, and letting her only child go on this trip was a big decision for her. Despite it all, my parents encouraged me to find a way. I didn't know how to make it happen, but I couldn't let money get in my way. I would never forgive myself if I missed this chance.

Over the next few months, I put all my energy into raising money for the trip. I ran bake sales and wrote for support to my state and federal government officials, local charities, and community groups. I had raised funds to attend the other trips, but this one was different; this time, I would sail on the ocean. After many months of hard work and help from friends and family, I raised the money. I couldn't contain my excitement. I was about to embark on what I knew would be a remarkable journey, but I never imagined how much it would change the course of my life. Looking back, I am so grateful I did not let money stop me. Money is an obstacle, but there are ways to overcome that hurdle if you love the goal and work hard enough.

The third ingredient to becoming the person you want to be is experience. Six months later, I traveled over 4,000 miles from Galt, California, to Reykjavik, Iceland, and got on a ship that would be my home for two and a half weeks. The ship was a classroom like no other; lifeboats hung off the side, water surrounded us, and there were no desks. I learned about glaciers, how changes to phytoplankton affect large animals, and how climate affects this ecosystem. A teacher showed us photos of whales and dolphins he had seen on other trips. One of the photos was of a blue whale, the largest animal on the planet. Like our nostrils, the blowhole of a blue whale is the size of an adult human head! We asked him about the chance of seeing a blue whale on this trip. He said there was less than a 5 percent chance we would see one on this trip.

"Despite it all, my parents encouraged me to find a way. I didn't know how to make it happen, but I couldn't let money get in my way."

Later that week, I would find myself on the deck of the ship, staring at one of the most breathtaking sights I had ever seen. Although it was summer, the wind was blowing, the air felt fresh and cool, and sometimes small pieces of ice floated past the ship as we made our way through the Denmark Strait. As my teacher stared out at the horizon with his binoculars, he said excitedly, "No way! Could it be?" The joy on his face made me feel like I was watching a kid on Christmas Day. He pointed, and we saw two puffs of air and water above the surface. They were from the blowholes of a blue whale and its baby. We couldn't believe it! There we were, standing on the ship's deck, watching blue whales on my first Arctic trip. At that moment, I understood how unique and remarkable the Arctic is. We spent the next two hours following the whale and its baby. It did not matter that we were now going in the wrong direction and running late. This was what the trip was all about. We were sharing space with the largest animal on Earth in a place many people only dream of visiting. We experienced a part of the ocean's wonder and beauty with experts to answer all our questions. Later, we learned this was the first time anyone had seen the baby or calf (baby whales are called calves). At fourteen years old, I was on the cutting edge of scientific knowledge, and it changed my life.

Finding the right people is the fourth ingredient to becoming the person you want to be. I came home from that trip knowing I wanted to go back and understand the Arctic better. I had full support and encouragement from my friends and family and especially from my high school chemistry teacher, Mrs. Crawford. She was a great mentor to me. I stayed in touch with her after I graduated high school, and we became friends. I was not good at chemistry and swore I was done with it. She reminded me that I was a strong person who could do anything and encouraged me to believe in myself. She pushed me out of my comfort zone and forced me to learn more. I said high school chemistry was as far as I would go. Despite her encouragement, I lost belief in myself along the way, but being a science major in college meant I would have to study more chemistry anyway. Back then, in my college classes, if I had a good experience with chemistry, I would tell her about it. Her continuous encouragement made me a better scientist and a better person. I carried her reminders with me throughout college. Mrs. Crawford passed away in 2015. I miss her every day, but her words of support have stayed with me.

The final ingredient to becoming the person you want to be is to try everything that comes your way! With Mrs. Crawford's advice, I started college and returned to the Arctic on a scholarship. I could not contain my excitement! I was getting paid to work and learn about the place that had a lasting impact on me.

My project was on the importance of the Pacific walrus in Alaska. I worked with Dr. Thomas Litwin and talked to scientists on the US Coast Guard Cutter Healy, the latest and most advanced polar icebreaker in the US at the time. He interviewed the scientists on board and recorded those discussions. In his videos, I saw Dr. Jacqueline Grebmeier for the first time. As I interviewed scientists for my project, her name came up, and I thought nothing of it.

During my junior year of college, I took classes through a non-profit organization called the Sea Education Association (SEA). During one of the seminars, a teacher spoke about the Arctic and encouraged us to meet with him if we wanted to learn more. I took him up on his offer, and he suggested I look up Dr. Grebmeier. I started researching her, and her face looked familiar, but I couldn't place her. Finally, it hit me - she was interviewed on the ship! I watched Dr. Litwin's videos again and looked back in my notes; she popped up everywhere. I decided to reach out to her.

I emailed her and waited patiently for her reply. Months went by, and it was time for me to use the persistence I developed during my earlier fundraising efforts. I wanted to meet Dr. Grebmeier, and I knew I could make it happen with hard work and help from someone I knew. That help came from Tom, my advisor from the scholarship program. He was a great mentor of mine. I met with him and explained what I wanted to do. He put me in touch with Dr. Grebmeier; she answered within hours this time! I didn't give up. I got help from someone who believed in me, and it paid off.

"Find someone that supports your goals and never let the rest of the world tell you that you can't do something."

Over the next few months, Jackie and I emailed back and forth. Finally, in May 2013, I spoke with Jackie and Dr. Lee Cooper about their Arctic research lab. I was nervous as Jackie and Lee were considered the greatest experts in their field of study. What if my experiences and thoughts didn't match up well with theirs? Yet, with these doubts, I understood how believing in myself could push me forward. My skills had led me this far, and I had to trust that they were good enough. Within minutes of answering the phone, my fears washed away. They were easy to talk to and cared to get to know me. During the call, they offered me the chance to return to the Arctic. Lee was serving as chief scientist on an upcoming research cruise. There was an extra spot on the ship, so they invited me to come along to make sure this was still what I wanted to pursue. Once again, I was dreaming big, being persistent, and working hard to get where I wanted to be.

Three months later, I flew to the Aleutian Islands of Alaska, boarded the largest ship I had ever been on, and set off on my second trip to the Arctic. I spent the next three weeks learning how to collect and study ocean water and animals. We used special tools to collect water, tiny sea animals called zooplankton, and other critters from the seafloor. I saw my first basket star and various worms, clams, and shrimp-like creatures called amphipods.

During the trip, Jackie asked if I was interested in becoming her graduate student, and I said, "Yes!" I will never forget the magical feeling from my first trip of standing on the ship's deck in Iceland, watching the blue whales. That trip and the people who encouraged me led me to this extraordinary opportunity.

Since I started working with Jackie, I have been part of 23 Arctic research cruises on four different ships from the United States and Canada, including one trip all the way to the North Pole! On the research cruises, my work consists of water and animal collections at all hours of the day. I have met and sailed with national and international scientists, many of whom have become people I know I can rely on. Sometimes, during cruises, we only sleep two hours a day, but every minute is worth it when I look out and see breaching humpback whales, walruses, and all the cool critters we collect from the seafloor. When I am not in the field, each day looks a little different. Some days, I review the data we collected or write about what we have found; other days, I look into a microscope for hours, identifying the animals we gathered on the ship. I completed my master's degree in 2016 by studying how acid in the ocean affects animals that make shells, in this case, clams. While doing my research, I spent my days keeping Arctic clams alive in a cold room in Maryland. Since Maryland has a very different climate, we had to learn how to keep the room at the correct temperature so the clams would live. I also organize supplies for the research cruises and write about our research to get donations to keep working. Just like the fundraising I did on my early trips! I finished my PhD, one of the highest levels of college degrees, in 2021, and since then, I have used my passion for the Arctic and the sea to teach students in Iceland, and now a little closer to home in Maryland. I spend my days inspiring college students and hope to have the same impact on their lives that so many people in this story had on mine. I have learned many new skills and continue to build my life as an environmental scientist and teacher. I want to leave you with one final ingredient that combines all the previous ones: Take every chance you get and embrace what might seem crazy!

Trip to the North Pole

C. GOETHEL

"Take every chance you get and embrace what might seem crazy!"

During my 8-week trip back from the North Pole, I discovered my dad had passed away. He was my hero and encouraged me to go to the Arctic for the first time. I still had three weeks left on my trip when I received the sad news. My dad struggled with health complications throughout his life, but this was sudden. A large part of me wanted to get off the ship and go home, but I stayed. I always promised him if anything happened, I would finish whatever I was doing, no matter how crazy it seemed. He encouraged my crazy ideas, and my life is richer and fuller because of it. I honored his lifelong, unending support for me by staying on the ship. When I returned from the trip, I was awarded the US Coast Guard Arctic Service Medal. What an honor for my hard work!

Jackie and Lee joined me on that trip, and their constant support and motivation helped me finish those last three weeks. They also made the journey a lot more enjoyable for me. Certain special people can truly make you who you are and give you the confidence to continue. Keep them, cherish them, and honor them by throwing yourself into what you love.

Life can be challenging. So many of my happiest and saddest moments have been out at sea doing what I love. But by mixing the right ingredients, time, and effort, you can overcome any problem. I waited eight years between my first and second journeys to the Arctic. However, I have made it to the Arctic at least once every summer since 2013. I have completed my graduate research and gained a lifelong mentor, colleague, and friend in Jackie.

CO-FOUNDER AND CHIEF OPERATING OFFICER

LOLA ADEYEMI

I started a business that no one thought they needed. I left my family in one country and started a business thousands of miles away in another from scratch. I returned to my home country of Nigeria to launch Magna Carta Health. It is a growing company providing healthcare to about 20,000 patients a year. Nigeria is a country my family and friends did not want to return to, and millions of people have fled. I got my motivation from death, lots of death. Disease and cancer robbed me of my brother, father, uncles, aunty, cousins, and countless friends.

The sad news about my brother's and father's passing crept in and spoiled our birthdays. It tainted holidays and cast a shadow on days that were meant to be happy; days that should be celebrations were joyless. I came face-to-face with cancer as I watched my father slip away. Seeing friends and family pass on due to disease forced me to try and understand what could have been done to prevent their deaths. These losses ignited my desire to change how people view their health, and they are at the very center of my story.

"I was tenacious in my resolve to complete things that I started, so if you were looking for someone to get something done. That would be me."

I had a very interesting childhood. I lived on three continents by the time I was 10. I was often sick and had health problems for many years. I was seen as "weak" because I lacked strength and stamina. I tried to make up for it by studying hard and doing my best at school. My suffering helped me solidify the belief that our lives are meant to help others. From a young age, even before my father was sick, I felt a need to help others. I saw this firsthand as my parents used their endlessly giving spirits to help their friends and family. I could not stop thinking about Mother Theresa and the amazing things she did, and when I saw the effects of the disease in my family, I thought, "What better way to help others than to become a doctor?" So, I became Dr. Lola and found out many of our health problems could be prevented by patient education and access to healthcare. A repeating cycle of not understanding how to stay healthy, being poor, and getting sick again has caused a health emergency. What people don't know can kill them. There is a saying: "If you think education is expensive, try ignorance." People are getting sick from preventable diseases due to a lack of education and information. The problem does not start with ignorance; it starts with poverty. Those most affected by preventable diseases are people experiencing poverty. Being poor makes it hard to get good healthcare and education. I'm a doctor, so I understand how diseases work, but I'm also a person who has experienced the sadness of losing someone. This motivated me to get a master's degree in public health at Johns Hopkins University. This taught me how much our health is linked to our environment. For example, there are over 100 illnesses linked to our environment, like chronic breathing problems from air pollution. Our environment affects our quality of life and our years of healthy living. Globally, 20 – 25% of deaths are caused by environmental factors that can be stopped or avoided. I continued to learn, train, and work, but I still yearned for something else, something more.

I decided to go to Harvard University for my master's degree in sustainability and environmental management. This filled gaps in my knowledge so I could move on to bigger ideas and concepts for the health of my patients. It satisfied my yearnings and gave me the resources to go forth and make a difference by founding Magna Carta Health. Due to my achievements, I was approached by companies offering great salaries. I gave up those opportunities to make zero; worse, I spent all my hard-earned savings to survive. I chose the more complicated, more challenging way to do things. I decided to leave the comfort and stability of the business world to assist people in the best way I could by providing personalized and preventive care.

Research has shown that the top five killer diseases (heart disease, stroke, difficulty breathing, infections of the lungs, and cancer) are preventable. Educating people and giving them access to preventive health care reduces their chances of getting sick. Focusing on their health before they get sick improves the quality of their lives and the lives of whole populations. My vision is to use today's newer technology to improve patient outcomes and the quality of life in communities to advance global health equity. Global health equity is a world in which everyone can achieve the highest attainable level of health, no matter their socioeconomic, demographic, or physical attributes.

Two months after my father passed away, my husband and I started Magna Carta Health. We didn't know how it would turn out, good or bad, successful or not. I had doubts and fears. I had never let fear stop me before, and I was not about to start now. I just knew we had to do it and take things as they came. Boy, did they come! We support the health of families and impact lives in ways that I never imagined. We run preventive health services to speed up diagnoses and treatments in poor areas of Lagos State, Nigeria. However, there is more work to be done. I am still learning and facing successes and challenges along the way.

One of the worst experiences was when I had to tell a dear friend that she was HIV positive over a video conference call. She was my ideal patient, "seemingly well," who came for a check-up. Unlike many of my stories, hers does not end in death. My friend is doing really well and managing her illness, all thanks to good preventive healthcare. Managing the health of patients can be difficult, but saving a life is a reward beyond measure.

We began Magna Carta Health with the simple idea that we could inspire others to live healthier by educating them on the benefits of preventative medicine. I knew one day I would run my own company. I didn't know when, where, or what I would do. You don't have to reinvent the wheel to make a difference in the world. You don't need to come up with the newest or most creative idea. You just need to have an idea that solves a problem you feel passionate about and believe in enough to do everything you can to make it a reality.

"You may feel, as I did, that you are meant to lead others with your vision of the future. I am proof that if you believe in yourself, stay focused, and don't give up, you can change the world."

FOUNDER AND EXECUTIVE DIRECTOR

ERIKA EBBEL ANGLE

My parents were immigrants and came to the United States with very little money and few possessions. With hard work and determination, they created a comfortable life for our family. When I was little, I accompanied my mom (a nurse, now retired) to free clinics and home visits, where she would work with members of the community who couldn't afford traditional healthcare. I saw firsthand that their children's clothes did not fit properly, and they had no toys to play with. I remember my mom telling me, "They have no toys—maybe you can give them some of yours." It's easy to take our situation for granted until we see others struggling.

Inspired by these experiences, I started volunteering at the local Martin Luther King Center in middle school. Once a week, I did homework help for students who couldn't afford formal tutoring and taught piano lessons another day of the week. I really loved volunteering. It felt great to help students learn that a subject or topic wasn't as hard as they thought and see their confidence grow. They were so happy; it was worth every second. I continued volunteering at the same center throughout high school. Middle school was also when I became interested in science.

"I started asking myself, 'What could I offer based on my skill set? I'm young, busy, and poor.' I decided to start a program to get children excited about STEM by having them do science fair projects similar to what inspired me."

When I was eleven, I had to choose between going to Washington, D.C., for a class field trip or to Cancun, Mexico, for a vacation because my parents couldn't afford both. I chose Cancun. Those who did not go on the class trip had to stay back at school with our English teacher, who gave us books like *Jurassic Park* and *The Andromeda Strain* to read. To accompany our reading, we did simple experiments such as growing bacteria in petri dishes. I loved all of it!

During my vacation in Cancun, we visited a crocodile farm. I found out that when crocodiles are very badly hurt, they might turn upside down, go into a deep sleep, and eventually pass away. That really affected me for some reason, and when it was time to work on a science fair project, I remembered this fascinating fact. I wondered if cells in the body also "commit suicide" when infected by viruses to avoid a horrible death. From there, my sixth-grade science fair idea was born. I knew I would have to find somebody to help me test this idea, so I pulled out the telephone book and looked up the phone numbers of local biotechnology companies. I called them and said I was an eleven-year-old student looking for a mentor to assist me with a science fair project idea. Most of the companies never called back.

Finally, I found Michael, the director of a local public health laboratory. He agreed to meet with me. He asked what I knew about microbiology (the study of organisms that cannot be seen with the naked eye) and immunology (the study of how the human body defends itself against infections and disease-causing organisms). As an eleven-year-old, I knew very little, so he sent me home with a stack of textbooks. I felt scared, nervous, and concerned that the material might be too hard to understand and over my head. I decided to give it a try anyway. Michael was willing to continue spending time to teach me, so I kept going back to learn more.

I decided I wouldn't be afraid of asking "dumb" questions. I would keep working with him until I fully understood the material. He was a wonderful mentor who was very patient, never became frustrated with my lack of knowledge, and always encouraged me to ask more questions. Finally, I told him about my idea for an experiment to see if cells commit suicide. He provided suggestions for how to do the experiment, leaving me to design the methods on my own. I infected some cells in Michael's lab with the least dangerous virus, the virus that causes cold sores on the lips (herpes simplex type 1). For hours, I waited and watched the cells to see what had happened.

Since I had never conducted this type of experiment before, I did not know what to expect. I learned there was no simple way to tell whether the cells were dying because they were being infected by viruses or because they were committing suicide. There was no way to "ask" them or to "see" this with an ordinary microscope.

The following year, I was sick in bed with the flu, and my dad gave me a book about Russian folk medicine. It described many herbal therapies and medicines for treating the flu. I was intrigued and decided that for my next science fair project, I would test an herbal remedy on the herpes virus I had worked with the year before. I called Michael and proposed my idea. He advised me to do some research at the medical library first. There, I found a scientific journal article about similar experiments conducted by a scientist in China.

"Our school-wide science fair came, and I didn't win, but going through this experience made me love science even more. I was allowed to work with real viruses and treated like an adult. It was fantastic!"

Unfortunately, the article was in Chinese, and I could not read it; however, the summary was available in English. Although the method was not described, the names of the ten most effective herbs for treating the herpes virus were included. I purchased some of them at a local Chinese herbal store. I prepared the herbs and took them to the lab for testing. Michael encouraged me to design experiments to test the ingredients the herbs broke down into (plant extracts). Two of the herbs prevented the herpes virus from infecting cells. I decided to focus on one of them for future experiments because it had not been researched as much as the others.

These experiments triggered many others, and I spent the next six years trying to determine which plant compound was responsible for the protective properties of this herb. During this time, I was fortunate enough to work at labs in large companies like Genentech and Perkin Elmer Applied Biosystems and at research institutions like San Francisco State University and Stanford University. Additionally, I took part in many science fairs, winning the California State Science Fair three times, becoming an Intel Science Talent Search finalist, and winning the National Junior Science and Humanities Symposium. It was fair to say I was a huge science dork - and proud of it!

When I went off to MIT (where many science dorks end up), I was starting over. Volunteering had become a big part of my life, and I needed a new program to become involved with. Despite many attempts, I couldn't find anything that inspired me. I asked myself, "What can I offer based on my skill set? I'm young, busy, and poor." I had recently read many articles about students' (especially girls) dwindling interest in STEM and couldn't understand why this was the case. Science has always been so incredible for me.

I was disappointed that many students thought it was boring. So, I started a program to get children excited about STEM by having them do science fair projects similar to what inspired me. I called a few law firms to see if they would help me set up my business idea called "Science from Scientists" (SfS) at no cost, for free. One agreed. Afterward, I had a small piece of paper saying I had a "business." Then, the hard work began. I launched a couple of science fairs at local elementary and middle schools. These schools and their teachers had never run science fairs, and they didn't have the materials or any plans to work with. I helped create a program plan, connected their students to scientists outside the school, and ran the science fair. Progress with the science fairs was slow because I was also a student at the time. In 2004, I became Miss Massachusetts in the Miss America program (see below) and decided that my platform would be getting children interested in STEM.

As Miss Massachusetts, I was invited to speak at many events attended by children of all ages. I talked about stereotypes and encouraged them to ignore conventionality and pursue their passions despite what they thought they were supposed to do and be. At the events, when I asked the younger children, "Who likes STEM?" all of them would raise their hands and excitedly say how much they loved science. However, when I posed that same question to high schoolers, most rolled their eyes at me. I would ask their teachers, "Why do you think this is happening?" and routinely, the answers would be: 1) We don't have time to do fun STEM in class, 2) We don't have much experience teaching STEM. Many elementary and middle school teachers, through no fault of their own, don't have strong backgrounds in STEM. 3) Due to testing requirements (for English and math, for example), we just don't prioritize science. So, I changed the SfS program to target elementary and middle school students. Research shows these years are a critical time to reach kids before deciding what they like and don't like. The program would need to run during school hours so that every child could participate, not just those who already loved STEM or whose parents could afford to put them in a STEM activity. It must also fit teachers' needs and align with state rules and educators. Our staff needed to be outstanding. We are very picky about the credentials of who we hire, as well as the staff's training and the content and format of the lesson plans. We did all of this. Things didn't go perfectly at first, but as time passed and we tried different approaches, we figured it out – as measured by positive feedback from the schools and teachers.

Today, SfS sends charismatic scientists into classrooms (during the school day) to teach science to students in grades three through eight. We have raised test scores and improved student interest in STEM by more than 89 percent (as reported by teachers). We are constantly looking for ways to improve. Our in-school programs teach 9,300 students at seventy-seven schools in California, Massachusetts, and Minnesota.

In 2017, we were the official educational partner of the America's Cup in Bermuda, reaching thousands of children and families and thousands more Bermudian children at the America's Cup Village. We also had a stage show, the SpectacuLAB, at Walt Disney World® Resort from November 2017 to January 2019. These programs brought science lessons to more than 100,000 additional children in the last year alone. SfS has more than sixty scientists working around the country and offers science lessons for school vacation programs and scientific training for teachers. Our staff is amazing. Without them, nothing would be possible. My job is to provide direction to SfS, help the team raise funds, build important relationships and partnerships (like with Walt Disney World® Resort or the America's Cup), and help as needed. I love my job because I meet so many amazing people and continue to bring my love of science to thousands of children and teachers around the country every day.

Erika and Mickey Mouse

MY MISS MASSACHUSETTS STORY

As a student at MIT, I clearly remember the events of September 11th and the horror everyone felt. I didn't own a TV then—I was never much of a TV watcher other than *Jeopardy!* and *Star Trek*. But after 9/11, my parents encouraged me to watch and read the news more. So, I bought a TV tuner for my computer to stay current with what was happening.

One night, not long after those events, I was watching TV with a group of friends. We were flipping channels and came upon the Miss America beauty contest. My friends turned to me and said, "You should do this! You have a talent. You have Science from Scientists (it was quite small then, but it did exist), and you always complain we don't have a chance to dress up enough!" I refused, saying that I didn't want to do it—but truthfully, I was scared to try. I was used to doing science in a lab, not being in pageants. I was afraid of failing. I refused again, but they kept pushing. Finally, they signed me up without telling me. I received a note from a local pageant director (in Massachusetts, you need to qualify for the state pageant by winning a local pageant) saying, "Congratulations! You've been signed up to participate." I knew my friends wouldn't let this go, so after much complaining, I decided to try it.

Truthfully, I expected the experience to be "stupid." I arrived at the pageant, bringing my only black business suit and my only one-piece Speedo® swimsuit (you get the picture here). I had never felt more out of place. I did not expect to meet many intelligent women. However, I found something very different. All of the contestants were incredibly well put together and sitting confidently in the auditorium. I was scared. By some stroke of "luck," I became contestant number one (the first one to audition). I had no idea what I was doing; it was all rather funny.

I vividly remember walking into my interview, standing there, and staring at the judges. They stared back. Finally, someone spoke. "Are you going to say anything?" I explained that I was new and had no idea what to do. They were very kind and gracious and helped me through the process. I marched my way through the day, and, of course, I didn't win.

Afterward, the friends who signed me up encouraged me to try again. I didn't want to, but they tried to convince me that this was like any other skill I needed to practice to learn. Additionally, one of the Miss Massachusetts pageant board members came up to me and said, "I believe you can be Miss America." I wasn't sure, but this support made me feel more confident.

That night, I called home and told my parents how it went. I explained that I didn't think I would do it again, but my father stopped me. He said, "Is there anything you can learn from this experience?" Truthfully, there were many things. I wasn't a great public speaker and didn't walk gracefully. I was pretty rough around the edges. He encouraged me to use pageants to improve myself and learn the softer skills necessary to interact with people socially. He was absolutely right. I signed up for Toastmasters to become a better public speaker. I hired a personal trainer to help get me in shape. I started learning about clothes, shoes, posture, and presentation.

It was quite a journey. I signed up for another local pageant and won. In my first year, I went to the state competition and placed second runner-up. It took three years before I finally won. During those years, I practiced all of the skills mentioned above. In my second year of competition, I found myself on the stage at Miss Massachusetts, standing next to the contestant who would go on to win the title that year. I remember thinking (and feeling a little hopeful but unsure), "I COULD ACTUALLY WIN THIS THING!" then, seconds later, the other woman won. One of the most exciting things I learned was the importance of believing great things can happen and that you can make them happen. During my third year, I changed my thought process.

Taking part in pageants was an important part of my journey because it shows how much I value working on myself and getting better at specific things. It's a reminder of how focused efforts to improve yourself can make a big difference. Pageants weren't a hobby but a part of my growth and development.

Not all women will do pageants, but it is important to identify areas in your life where you can improve your skills to achieve something great. It is in your hands and under your control. We typically don't like to identify our weaknesses because they make us feel bad, but you can't become a better human being if you don't focus on self-improvement. Don't be embarrassed about your shortcomings. We all have them. Don't hide them or lie about them. Bring them front and center and figure out how to improve them, or find someone to help you.

Imagine yourself as being born holding a "hand" of cards that range from deuces (the worst) to aces (the best). Imagine that these cards represent your skills. You aren't born with all aces, but you can slowly affect the course of your life to upgrade your deuces into aces. It takes time and effort, but you can do it. Then, you can play the game of life with the best hand possible!

MEDICAL EPIDEMIOLOGIST/ ADJUNCT PROFESSOR

EVA SUARTHANA

"What I love about science is that it brings people together for the betterment of human life and humanity. I hope to serve as a mentor for up-and-coming scientists, as others have done for me."

My mother jokes about my early years, "You fell sick so often that you smelled like medicine." In junior high school, one of my sisters broke her leg; the other broke her arm, and a famous surgeon fixed it. I found my inspiration! I wanted to help sick people get better. I wanted to become a medical doctor. I was the valedictorian of my high school class (the student with the highest grades). This meant I could study at the Faculty of Medicine at the University of Indonesia (FMUI) without taking the entrance exam. I felt very fortunate as FMUI is Indonesia's oldest and best medical school.

In medical school, I worked on a project to teach young mothers in a poor community how to make a salt and sugar drink to prevent dehydration (not getting enough water) after having diarrhea. It was an eye-opening experience. It showed me that working in the community can be more important than treating people in the doctor's office. I could teach people how to stay healthy and prevent them from getting sick. Also, during this project, we found that if companies provided safety training for workers, they could prevent some work-related sicknesses. These experiences are at the root of my passion for teaching, research, and working in public health.

"People say I was fearless to move to a new country alone, without knowing anyone. My father's adventurous blood is in me, and my young spirit was excited to travel and explore the world."

After graduation, I worked in the Department of Community Medicine of FMUI, supervising medical students and teaching family medicine. While there, I took an international class on epidemiology given by professors from the Netherlands. Epidemiology is a field of work that looks for answers about the spread, patterns, and source of a disease or health issue and how to prevent and control it. I interviewed one of the professors, Dr. Diederick Grobbee, who left his medical practice to do research. I always thought medical doctors practiced in clinics or hospitals to treat patients. He had no regrets because he felt he had a greater impact on patients through research. He also became my inspiration!

"I learned that if you keep trying hard and don't give up, you can end up in some incredible and unexpected places."

I was 25 when I won a scholarship for a master's degree in Clinical Epidemiology in Rotterdam, the Netherlands. Moving to Rotterdam was a real adventure for a girl born and raised in Indonesia. I felt worried, nervous, and excited at the same time. I did not come from a wealthy family, so studying in another country never crossed my mind. It was harder on my parents. I had never been separated from them, except for out-of-town assignments in medical school. My mother is a housewife. She is an intelligent woman but couldn't complete high school because her family did not have enough money. She had never traveled outside of Indonesia, so the idea of sending me thousands of miles away was frightening. My father was an adventurer and knew this chance might not come again. So, he convinced my mother to let me go, and they gave me their blessing to leave. When my father was very young, he left his family in a small village in Bali and started a life in Jakarta. Jakarta is the capital city of Indonesia. People come from all over the country to live in Jakarta for better education and jobs. He started as a worker in the purchasing department of a hotel. He completed his bachelor's degree in English while working there to get promoted. He became Purchasing Manager at many five-star hotels before retiring almost ten years later. He often worked with expatriates (people who live outside their native country) from around the world, just like I was going to become.

In Rotterdam, I got used to a new lifestyle, learned new skills, and made friends with many students just like myself. The classes I took were in English, which was a big help for me because I didn't know how to speak the native Dutch language. Also, since I am a practicing Muslim who grew up in the largest Muslim country, I had to learn to live as a minority and how to deal with looking like a minority. Most Indonesians are considered moderate Muslims and are very peaceful. I was not wearing the headscarf then, but my friends knew I was a Muslim. I soon learned that the way the Western media portrayed Islam (especially after the September 11th attacks in the United States (9/11)) was far from the Islam I knew and practiced. After a few months, I decided to wear the headscarf.

"I wanted to show my identity as a Muslim so people could see Muslims as ordinary people just like them. It motivated me to do my best in my classes and be kind to everyone."

My parents were worried people would treat me unfairly. However, my friends all reacted positively. "Your headscarf doesn't change who you are," they said. My boss said, "In my opinion, as long as you don't cover your face, you won't have any problem because people can still recognize you. It's your right to practice your belief." I felt a huge relief because I could show people the true colors of Islam, and they treated me with respect and fairness.

After completing my master's degree, I got the opportunity to do a PhD program at Utrecht University in the Netherlands, supervised by Drs. Diederick Grobbee, Dick Heederik, and Evert Meijer in the epidemiology of lung diseases. I learned that medical prediction studies are like weather forecasting. Data is used to predict the chance of rain, snow, and storms so people can take action. Since some medical tests and treatments are expensive and risky, prediction studies help us understand how likely it is for someone to get a disease or become ill. Prediction studies also help doctors choose tests and treatments that are the safest and most useful. It is fascinating!

I made a prediction model (like a math equation) to see which people working in bakeries would get sick. Breathing in flour can cause some workers to have asthma. I never met the workers, but I used their health information and computer programs to predict their chances of getting sick. I developed an easy-to-use checklist to help doctors detect sick workers sooner. Our model is used as an example by the European Respiratory Society Task Force for the early detection of work-related diseases. What an honor! It is still used as a screening tool for Dutch bakers. I am delighted that my work can improve workers' health on a large scale.

While doing my PhD, I won a scholarship in Canada, where I would learn about work-related asthma from experts, including Dr. Denyse Gautrin, who was also an epidemiologist. One of the perks of doing a master's or PhD degree is sharing your work at scientific meetings with experts in the audience. While attending the American Thoracic Society annual meeting to show my research with Dr. Gautrin, I attended a women's dinner and met female researchers from across the US. I was motivated and inspired by their stories, just like I hope my story will do for you! The women were very genuine and offered advice. It was my first time seeing researchers of all ages spend time together and share lessons learned. I met Dr. Kathleen (Kay) Kreiss from the National Institute for Occupational Safety and Health (NIOSH) there. She knew of a job where I could continue my research and gain more skills. She suggested I apply to the Epidemic Intelligence Service (EIS). EIS officers are known as the detectives for "outbreak investigations" across the US. They study diseases that happen more often in certain groups of people, regions, or seasons of the year. I was thrilled! Never in a million years had I dreamed of a job at NIOSH. It is one of the biggest government groups helping to prevent work-related sickness in the US.

A year after I completed my PhD, I got accepted to the EIS and mentored by Dr. Kreiss! I was one of just a few international experts accepted into the program. This was an excellent chance for me to improve my skills, do more research, and connect with experts about diseases and conditions related to the workplace.

While at the EIS, I was pregnant and gave birth to my first child. It was life-changing and tough because our families live in Indonesia. Fortunately, my husband is a nurse, so he cared for the baby and me. I slowly returned to work by the time my daughter was six weeks old. We did not want to put her in daycare, so my husband became a stay-at-home dad. I learned to juggle my new life as a working mom. Now, I have two daughters and a son. Balancing family life with my work is challenging but satisfying.

After completing the EIS Program, I accepted a job from Hôpital du Sacré-Coeur de Montréal as a full-time researcher with my former mentor, Dr. Gautrin. I began to teach and guide students who were doing their research and wrote requests to get money for my research projects. I developed prediction models to find workers with asthma due to dust, fumes, or chemicals at work. The best test for this is only available in a few places in the world. So, getting a diagnosis is complicated and takes too much time. We created a prediction model to fix these problems and used it to make online calculators and mobile apps. This helped to spread its use globally. The project received several awards because it was creative and could help improve workers' health. It was recognized through a famous grant awarded by the Canadian Institute for Health Research as a valuable tool made possible through international collaborations.

My work in women's health in Canada started when Dr. Togas Tulandi asked me to be an Adjunct Professor at McGill University. I am part of a team that supports research for medical doctors studying to become specialists in obstetrics and gynecology (doctors who help pregnant women give birth). I teach epidemiology and statistics (math concepts used to evaluate research data). I love my job helping doctors design research studies and teaching them how to use data to answer medical questions. Many of them come from around the world. We talk about a variety of different health topics, like difficulty getting pregnant and even cancer.

Previously, I was part of a nationwide project to raise awareness of cervical cancer and create a program for patients in poor areas to get tested and treated. Cervical cancer is the most common cancer in Indonesian women. Most women are diagnosed late, so outcomes are poor. We quickly learned we needed different ways of running the program for each province due to differences in cultural values. It was challenging and required the efforts of many talented people and money management skills. We succeeded by creating a positive work environment and becoming team players. Everyone was working towards the same goals. I am happy knowing I added to the screening and medical care system used to help people; it was very gratifying. I could not have done this without Dr. Setyawati Budiningsih as my mentor.

Now, I work full-time in the Health Technology Assessment Unit of the McGill University Health Center. Our team, led by a senior epidemiologist, Dr. Nisha Almeida, uses effectiveness (how well things work), safety, and cost information to help the hospital decide when to use new equipment, drugs, and procedures. I learned to look over all the information I have carefully. This helps me provide accurate and reliable advice to doctors and hospitals. This work has broadened my view of how research impacts patients' lives and the government rules for treating patients. I love science because it brings people together to better human life and humanity. I will continue teaching and doing research, as education is the key to strengthening our communities and the younger generations.

People say I was fearless to move to a new country alone and without knowing anyone. My father's adventurous blood is in me, and my young spirit was excited to travel and explore the world. Today, we study, work, and contribute globally and enjoy exciting opportunities to work with others who share and expand our scientific and other interests. Confidence, hard work, motivation, and teamwork are essential to success.

"I will continue teaching and doing research, as education is the key to strengthening our communities and the younger generations."

CONSULTANT

ANNE CAMILLE TALLEY

"Look for the confluence of your interests to find opportunity."

In fourth grade, I was a bit "behind expectations" in learning multiplication tables, so it was time for some tutoring. My next older brother, Patrick, a math wizard, was assigned to help. He quickly had enough of my complaints: "It's too much!" and "I can't remember all those numbers!" He advised: "There's no reason to put off what's going to happen anyway; neither of us is getting out of this until you learn them all. So, pick one, any one, to memorize first, and we'll start there." I picked 7's, and what Pat did next surprised me. He got me started with the sequence I had just memorized: 7, 14, 21, 28, 35… then he showed me other ways to achieve the same results. For example, adding seven plus seven is always the same as multiplying seven times two. Similarly, seven times three is always the same as seven plus seven plus seven. I already knew how to add, so I had a "work-around" in case I forgot a number in my memorized "7-times table." Hooray! He showed me that when I struggle with one discipline (remembering all those numbers), I can find help from old friends (my addition skills) and clever tricks (add a zero, and Bang!... you have ten times any number). I had done it all in my head without stopping to pull out a calculator, a book, or even a pencil! Where addition and multiplication come together (a confluence), there's a magical fix for: "I can't remember all those numbers!"

At about that same time, my mom sent me off to Saturday morning art classes in the basement workroom of the convent (house for a community under religious vows) where the nuns who taught at my school lived. Those were light-hearted lessons in drawing, image composition, watercolor painting, and, now that I look back on it, the joy of knowing my teachers outside the classroom, in their homes.

Every Saturday morning, the art students and the nuns spent time together at the place where formal and informal things came together, learning and playing, and we were finding our creativity. I didn't realize until much later how much I learned about science in those art classes. I rode my bike a mile back and forth to the convent on those early mornings and, over the months, built up the skill to ride hands-free, navigating the turns by leaning slightly left or right. My mom never knew I was studying physics like that!

Ten years after completing a bachelor's degree in biology, I worked as a junior technician in a university laboratory. Around the lab bench, as we ate lunch together, my co-workers told incredible stories about the work of a prestigious marine biological laboratory. I also heard their warning: "To work there, you have to be in the right place, at the right time!" I complained to my dad that going there for my next job sounded great, but I didn't know how to figure out "the right place" or the "right time." His advice: "You'll never be in the right place at the right time if you don't put yourself there at some time. Go, visit, and ask who and what they're looking for."

His idea was to find an intersection of place and time by choosing a point in the stream of time and going to that place to look for opportunity. I took his advice and became the laboratory's first public relations officer. At first, the laboratory director was concerned that my biology degree and experience might not have prepared me for science administration. He asked one of my job references, an English professor, if he thought I had the skills to imagine something that had not existed before (like a public relations effort for the laboratory) and bring it to life. My professor said he had seen me pull together a wide range of facts into meaningful ideas and stories many times and that I had strong enough relationship-building skills to ask for help when I needed it. That sealed the deal.

A few years later, Patrick completed his master's degree in business administration (MBA) and was going to work for a giant computer company.

The day he graduated, he looked at me and said, "It doesn't appear that some wealthy prince is going to come along and whisk you off on his white stallion to a life of luxury. Why don't you get an MBA, too? That way, with your biology degree, you could translate science for business people and business for scientists… and always find a job to support yourself." Four years later, I had my MBA, and, as it happened, a new word had just come into use: biotechnology. In that field, people were using scientific discoveries to create products that people needed (like at-home blood pressure cuffs) and building businesses out of those products.

"…[I] uncover insights that scientists and engineers use to create products that help patients get and stay healthy."

I was one of only two people I knew with degrees in both science and business. That person and I started a consulting business, advising companies that wanted to invest in biotechnology but didn't understand the science. We were translating science for business people at the confluence of both fields. I settled into the field of market research for biotechnology and pharmaceutical companies, where I got into the "flow" of information streams about what customers or patients need. I uncover insights that scientists and engineers use to create products that help patients get and stay healthy. This is yet another way for me to translate my diverse background into an exciting career. I've crafted a life where I look forward to going to work as soon as my feet hit the floor every morning.

Here's something I learned along the way about technology. I learned that I could "buckle down" in the way Patrick got me started with my times tables. I could learn just enough about each new wave of tech to use those new tools. I didn't have to become an expert in every new wave. I learned I was happier when I got my skills sharp enough to use them just a bit… while still calling on "old friends" and clever tricks to continue to find intersections of my skills and new opportunities. When it's time to "buckle down" like that, I'm reminded that the harder you work, the more opportunities you get.

Science, Technology, Engineering, Math (STEM). Add the religion my parents gave me, using it to grow my hope and belief. Then, mix in Art that feeds and expresses beauty and creativity. Rearrange STEM plus R, plus A, and I've been playing in a STREAM, stepping from stone to stone, watching for confluences and opportunities all my life and finding them.

Oh, and my "prince" found me, too. That's a story for another memoir.

DIRECTOR OF SOLUTIONS CONSULTING

MELANIE L. FLORES

"I had to learn by doing, with limited resources, and in the heat of the moment."

Growing up, I learned to thrive in the soil in which I was planted. My dad served in the US Navy, and we moved five times between my 2nd and 9th grades of school. Adjusting to so many new situations, schools, and peer groups taught me to be resilient. Being the "new girl" makes you feel vulnerable. The one place I always felt strong and capable was in my STEM classes. I was good at math and science, and throughout middle school and high school, I was blessed with teachers whose passion and excitement for STEM rubbed off on me. While learning about microscopes in middle school, I saw a tiny living thing with just one cell called an amoeba. We were learning about amoebas that very week in class. My biology teacher, Mr. Perrenot, was so excited he fist-bumped the air! I remember feeling like a real scientist and my chest swelling with pride.

Teachers like him helped me appreciate that science could be thrilling. My Advanced Placement Physics teacher, Mr. Dickens, encouraged me to apply to the Massachusetts Institute of Technology (MIT) for college. Many months later, I was excited to find a fat envelope from the admissions office in my mailbox! I was thrilled, and I agreed to the offer.

"Observation and persistence are powerful problem-solving tools."

I finished my studies there with a degree in chemical engineering. Chemical engineers design procedures to make, change, and move different kinds of materials. They make almost all the products we use every day, like paper, food, drinks, medicines, fertilizers, dyes, and household products.

After college, I worked as a Production Shift Supervisor at an optical fiber factory in North Carolina. Optical fibers are most often used to carry light for communications. They can do it over longer distances and at higher rates than old-fashioned copper cables. Optical fibers are also used for lighting and medical imaging because they can carry light into or create images of small spaces in the human body.

The factories where the fibers are made have many different departments, just like a department store. The factory staff work in set time blocks, typically 8- or 12-hour shifts. It was my responsibility to ensure my department ran safely and effectively and made the expected amount of product. I led a crew of 24 operators and one technician, who ran the machines that turned glass into optical fibers. The machines ran at very high temperatures and speeds.

The factory ran all the time, every day. Sometimes, I worked from 7 in the evening until 7 in the morning. Other times, I worked the opposite, from 7 in the morning until 7 in the evening. This schedule is known as *rotating shift work*. It is exhausting because your body constantly has to change rhythms, and you live on a vastly different schedule from most people. Imagine going to work at 7 p.m. on a Friday night when most people are just starting to enjoy the weekend! The factory was constantly changing, which sometimes made people tense. Still, it was a great place to learn a lot quickly. There were many on-the-job lessons not covered in my college textbooks and lectures.

First, I learned how to decide which problems are most important. In a workplace that never sleeps, whether in an emergency room at a hospital or a factory like mine, you need to think clearly and constantly reset priorities. If something went wrong at 3 a.m., I had to resolve it because I was the supervisor. I had to learn by doing, with limited resources, and in the heat of the moment. For example, when many machines quit working all at once, I had to develop a plan to fix them and understand enough about how things worked to know when to call for help. As I gained experience, I became better at my decision-making.

Second, I gained priceless experience learning to work with people from all walks of life. Many of my employees were twice my age. Some of them did not take me seriously. I had to earn their respect by spending time on the factory floor with them, learning the ins and outs of their job. At first, one of my top workers seemed distant towards me.

As I got to know him, I realized he was very good at troubleshooting problems (fixing and figuring out what was wrong), so I started turning to him when problems arose. He appreciated my confidence in him and became a trusted partner.

"Anything of permanence that I can help grow from the ground up captivates me."

Third, I figured out how to talk and write briefly and clearly. When my shift was over, I had to clearly describe what happened during my shift and any open issues to the incoming supervisor; this is called the "pass down." One example of a pass down would be: "Furnace 25 is down for scheduled maintenance. Furnace 30 has been repaired and is back up and running. For some reason, Furnace 40 is not heating up properly, and we are still troubleshooting it."

Clear communication is vital to the people you call for support, such as maintenance or specialized engineers, so they can help diagnose problems quickly. If your message is unclear, precious time might be wasted. It is also critical to be aware of products entering and leaving your area because, like a clogged sink or a traffic jam, a pileup of products in one area is a problem that needs fixing.

After a year in that job, I transferred from the production floor to engineering. My new job was to find ways to increase our outputs ("yields"), cut costs, reduce downtime (time that a machine is not running), and improve safety. Since the plant operated around the clock, when they had issues, I would get a call no matter what time it was. Sometimes, I'd have to get out of bed in the middle of the night and come to work to fix the problem.

As a manufacturing engineer, I learned many vital lessons:

- Communicate your work clearly and in writing. Otherwise, I could get a call at 1 a.m. that starts with: "What are we supposed to do again?" or "Something isn't right." These types of emergencies happen less often with proper planning, training, and clear communication.
- Work carefully and follow a step-by-step approach. Test changes on a small scale, such as on ONE machine. If your change does not go well, it is much easier to fix the problem if it is just in one area. This approach is known as "scaling smartly."
- Observation and persistence are powerful problem-solving tools. I would sit in a chair next to a piece of equipment and watch it like a hawk to find clues as to why specific problems were happening.

After five years, I was picked to join the startup team for a new manufacturing plant. I had some doubts at first: what if I failed? Still, the fact that I was asked to do it meant that people believed in me. I decided I needed to believe in myself. Giddy with excitement and more than a little scared, I said yes to the challenge. Like my acceptance to MIT, I realized that starting up a new plant was a once-in-a-lifetime opportunity, and I was not about to pass it up. It did not disappoint.

After someone installed new equipment, we had to "debug it," meaning we had to perform practice runs and ensure the machines did what they were supposed to do. If they didn't work correctly, we had to figure out what was wrong and fix it. Sometimes, the solution was simple. For example, a loose wire might need to be tightened. Other times, it was more complicated. Occasionally, we had to take machines apart piece by piece to figure out the problem. Along with debugging the equipment, we had to hire hundreds of people, train them, and write procedures for how everything should function to get the factory up and running. We worked grueling hours for months on end. There were days I was at work until the wee hours of the morning. Our hard work paid off one summer night when we made our first optical fiber. The entire factory celebrated together. That moment will stick with me forever.

"In a workplace that never sleeps, whether in an emergency room at a hospital or a factory like mine, you need to think clearly and constantly reset priorities."

My superpower is my creativity, and I'm not alone. A research study found that 91% of 5th-12th-grade girls describe themselves as creative, and 72% say that a job that helps the world is vital to them. That makes STEM an excellent career fit for girls. You can be creative and help many others at the same time.

I landed that first job because I had a STEM background. While I'm no longer in manufacturing, that experience has been valuable to everything I've done since then! I use what I've learned about understanding the overall plan (the big picture), how to decide what's most important, collaborating with others, and speaking plainly. I've also grown wiser through education and experience with e-commerce. These skills continue to open doors for me decades later, and I will use them for the rest of my life.

Since I always had to adjust to being an outsider with people and places as a child, I look for opportunities to build new things for others. The bigger, deeper, and longer-lasting the impact is, the better! It helps me compensate for the

M. L. FLORES

constant uprooting I experienced as a child. Anything of permanence that I can help grow from the ground up captivates me. I'm thankful for the opportunity to help build something as lasting and impactful as an optical fiber manufacturing facility. My work helped bring faster internet to people all over the world!

CHIEF OF STAFF MARINE ECOLOGIST

TRIKA GERARD

I was born and raised in St. Thomas, US Virgin Islands, one of the United States territories in the Caribbean. Close to two million visitors travel to this American paradise every year to get an up-close view of the sandy beaches and clear blue waters and experience the Caribbean culture under year-round sunny skies. Snorkeling or building sand castles at the beach are some activities that visitors and locals enjoy. On any given day, you can spot iguanas feasting on lush green tree leaves in the hills, tropical flowers lining the streets, or kids climbing mango and coconut fruit trees for a quick snack. As a kid, I loved watching how the different living things around me interacted with each other. It was exciting to see how they connected and did things together. Little did I know this fascination and intrigue would become my passion and lead me down a path toward learning more about living organisms.

Throughout my early childhood education, I was drawn to the mystery and awesomeness of science. I arrived at high school eager to prepare for college and my career; however, there was one small obstacle before me: a tenth-grade biology class. No one wanted to take the biology course taught by one of the toughest educators to grace the halls of my high school. It was taught by Mr. Austin Gumbs, who is known for being strict but thorough. Everyone knew his course was one of the most difficult to pass. As luck would have it, I was assigned to his class, and he lived up to his reputation. He didn't like it when people were late and demanded each student be fully attentive and ready to learn during class. We were overwhelmed with so much information about living things, but he expected us to study hard and understand everything. As it turns out, I loved every minute of it! It was his biology class that affirmed my love of science. Mr. Gumbs encouraged students to look in their environment for examples of concepts taught in our textbooks. This practice led me to realize that I observed everything in great detail and desired not just to ask questions but to answer those questions! For example, I would look at the steamed fish on my dinner plate, observe the skin color pattern, and wonder how it was different from the steamed fish from last week's dinner.

"My interests were unpopular, but they were mine and worth pursuing!"

Mr. Gumbs's class made it clear to me that I thought like a scientist. The practice of asking and answering questions from detailed observations gave me a wealth of information and knowledge. This knowledge empowered me, and I decided to study biology in college. It was a moment of conflict for me. I had to make a decision that had a far-reaching impact on my life and career. If I had followed the advice of my friends, who warned me that passing his biology course was impossible, I might have missed the opportunity to realize my true calling. This moment may have marked the first time I was confronted with a challenge and learned how to overcome it. When it's your turn, you may pleasantly surprise yourself in the process. I dared to be different! While I was excited about my career choice, I felt alone and discouraged because what I wanted differed from what my friends and family wanted. No one in my family pursued a career in science. Conversations at family events were never science-related. The women in my family were business professionals, and my dad and uncles had military or law enforcement jobs. Cousins my age were interested in being educators or journalists. Friends and classmates hoped to become lawyers, school principals, and small business owners.

"The journey was sometimes challenging but worthwhile in the pursuit of a fulfilling career. If I had followed the advice of my friends...I may have missed the opportunity to realize my true calling."

On the other hand, I always wondered what marine life lived beneath the water at the beach or how the ocean around me was formed. Simply put, I was labeled a "nerd" or "geek" because my career goal was complicated to those around me and sometimes misunderstood. At first, I didn't follow what I loved because not many people liked it, and there weren't any role models to help me imagine myself in that career. So, I applied to college as a computer science major instead. Then, when I registered for my first semester classes, I changed my major to medical technology. After meeting folks in college with similar career ambitions, it became abundantly clear that my true calling was to study biology, so I switched my major again. My interests were unpopular, but they were mine and worth pursuing! I was motivated to find out the answers to many questions about the environment. Even though not many people supported my goals, my ambitions pushed me to work hard and achieve a doctorate (PhD) in environmental science. This journey was sometimes challenging, but I learned it is okay to be different in the pursuit of a fulfilling career. Marine environmental science is dominated by white males. As such, I rarely saw anyone who looked like me at scientific conferences, at work, or at colleges and universities. Most often, I was one of a small pool of women in the field and one of a few African Americans. This sometimes made me feel lonely and was somewhat discouraging because I felt I didn't belong. Despite this, there is no reason to feel pressure to change your goals because your choice is unpopular. If you are excited about your career choice and interests, embrace them and own them! Dare to be different!

As a scientist for the federal government, I have conducted research studies that are used to manage and conserve marine natural resources so that everyone may have access to and enjoy them. My research has helped us make meaningful choices, like rules around how and when to fish and when to close fishing areas. These choices can impact everyone in our country. My research has also brought about justice on behalf of the country when there has been an environmental infraction (breaking the law) or event, such as an oil spill. My twenty-plus-year career with the National Oceanic and Atmospheric Administration (NOAA) has afforded me endless opportunities to be a force in an industry that does not have many African American women. I currently hold a position as the Chief of Staff for Science and Research, but for 13 years, I focused on early marine life history (baby fish) studies.

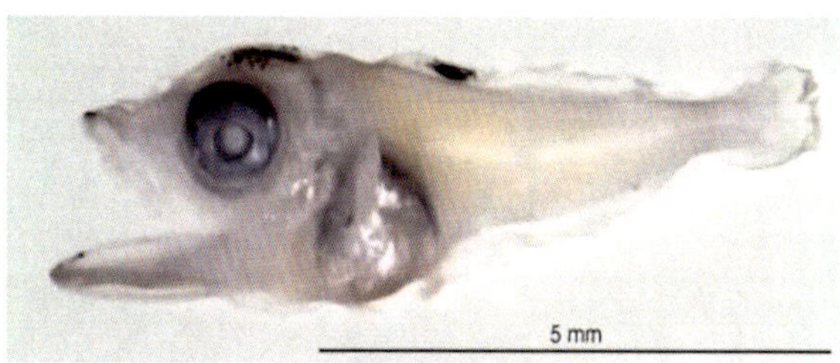

Baby fish depend on ocean currents to carry them before they can swim. While thousands of fish may hatch at a time, only about 1 percent survive. I studied fish to figure out where they were spawned (laid eggs) and how far the babies traveled in the ocean. This data is important because if something stops them from growing into adults, we won't have the luxury of enjoying them, commercially or recreationally. Therefore, we must understand their early life to help them grow into adults. Since fish are in all oceans but at different depths, I snorkel and scuba dive during research cruises and travel to fascinating places in the Gulf of Mexico and the Caribbean to collect them.

Another highlight of my research is traveling to scientific conferences worldwide to share my research results with other scientists. As thrilling as my job is, it's disappointing that I still don't see many people who look like me in this line of work. Less than one percent of African American females earn a doctorate and ever work in science or engineering. I am one of a handful of African American female scientists at my laboratory, and I have often experienced the same thing when traveling to science conferences. I hope my story will help change this!

Although I sometimes feel like a small fish in a large ocean, this reality drives me to be the best scientist possible. I recently celebrated the most noteworthy accomplishment of my career. I was awarded the NOAA Administrators Award for designing and executing the first fisheries oceanography survey around Cuba. A NOAA Administrator leads the organization, and the administrators' award recognizes staff (individually or in a group) that significantly contribute to NOAA programs in scientific research, among other areas. Nominations go before a review board where the contribution is assessed for importance, uniqueness, originality, or excellence in project planning and completion, among other factors.

In terms of government relations, dealing with Cuba has been challenging and seemed impossible to manage for decades. From a scientific perspective, Cuban waters were a great unknown. But, given the closeness to US ecosystems and the fact that marine wildlife does not know boundaries, it makes sense to include Cuba in our North American research to understand the whole ecosystem completely. As you can imagine, I was thrilled to be nominated and honored that the administrator deemed our research worthy of such an accolade. An achievement like this takes me back to fondly recalling that young girl who dared to be different for her love of science!

ASSISTANT CHIEF (RETIRED)

SANDY JO MACARTHUR

"Many people had certain expectations of me. I found it so odd that I was encouraged to do things I found boring."

When I was younger, I used to get a tremendous feeling of joy whenever I experienced something new. I remember thinking every experience was a gift, so I was always looking for the next one. That's why I loved to dream about doing things I read about and visiting places I saw on television. I was jealous of people with adventurous hobbies like mountain climbing, sailing, and scuba diving. I thought exploring the world under the sea or from the top of a mountain would be amazing. When I told my high school guidance counselor I wanted to learn to scuba dive and explore the ocean, he laughed and told me it was silly. Some of my teachers said I would make a great math teacher or that I should be a secretary. But I wanted something different. I wanted something more exciting, but I didn't know what just yet. I started believing my guidance counselors and was feeling very sad. Fortunately, I chose not to listen to them and listened to my parents instead. They were in the plumbing business, and at that time, it wasn't a very common field of work for a woman, my mom. Having a mom in a non-traditional job and watching her work inspired me to not place limits on myself or my abilities. She always told me I could do anything I wanted to in life, and I took that to mean I should follow my dreams.

My childhood friend, Marge, is the type of friend everyone hopes for, and fortunately, she is still in my life today. I always convinced her to try new things; she always had my back. Even though she thought I was crazy. We did everything together: school, babysitting, camping, and dreaming. And I told her everything: my fears, my first crush, and, most importantly, my big dreams. She thought my dreams were a bit too big, but she always cheered me on and supported me every step of the way.

One day, when I was fifteen, Marge and I were on one of our big biking adventures near the river behind her house. I found a newspaper and saw an ad for scuba-diving classes. I will never forget it! A way to make my dream come true was staring me in the face. You see, I lived in the Midwest (nowhere near a body of water deep enough to scuba dive), and the dream of scuba diving didn't seem possible. At the time, it truly was my number-one dream. The ad said: "You must be at least fifteen years old and an excellent swimmer. If that's you, sign up for an exciting hobby that takes you under the sea."

Since I was a junior lifeguard, I figured I didn't have to worry about my qualifications as a swimmer. When I showed it to Marge, she smiled and said, "Of course, you want to scuba dive. Go for it!" When I signed up for the course, I realized that all my math and science courses would come in handy. I learned to calculate how much nitrogen my body would take during each dive, letting me know how much time I needed to be on the surface between dives to be safe and how deep I could go. I learned about maximum dive times. The deeper one goes, the greater the water pressure. The greater the water pressure on your body, the more air gathers in your body tissues. This can cause a diver to become ill underwater or even after they surface. Also, our blood takes in nitrogen quicker in deep dives than in shallow dives. This can cause you to feel dizzy and even pass out, which would be very dangerous. So, it is essential that if you go deep, the dive must be very short so your body and brain can work properly. It was all new, exciting, and interesting, and it was an interesting way to learn science. It was also the beginning of my journey to follow my big dreams. Dreams that took me from the Midwest to California to becoming a police officer at a time when girls didn't do that sort of thing.

Sandy and Marge

I graduated from Arizona State University with a bachelor of science degree in criminal justice and decided to join the Los Angeles Police Department (LAPD). It sounded exciting, a bit risky, and very hard. I spent the next six months running and getting in the best shape of my life.

I wanted my big dream to include something that could make a difference in the world for others, so I joined the LAPD when not many young women thought of policing as a career. On day one of the academies, I realized what a wild and crazy dream I had decided to pursue. I had to be a strong athlete. I had to apply all of my studying skills to learn law, policy, and procedures. Unfortunately, I had to talk to many people who didn't believe girls could be police officers, too. I was told I didn't belong simply because of my gender. Some felt I would not be there in a fight, and they were unsure I could handle the stress of an intense situation on the job. Instead of getting angry about their prejudices, I decided to be curious and figure out why they felt that way. I asked questions to try to understand. I didn't take their attacks on me and my abilities personally. I knew I was strong and capable, and I was on my way to fulfilling my biggest and best dream of all!

The academy training was difficult and included studying textbooks, science, self-defense, physical fitness, and law. It was the most challenging six months of my life. When I graduated from the academy, I was one of only 196 women on a police force of over 7,000 officers. I felt a great sense of achievement, loved my new career in service, and realized I found a purpose in a life dedicated to serving others.

When I first joined the department, there were few young women and no female mentors to help guide me through this challenging career. I made some mistakes along the way, but I always felt deep inside that I was strong and could overcome anything that came my way.

I survived some pretty tricky and risky situations while cruising around the streets. I overcame bullying by male officers who did not think I belonged in the force. I decided that confidence in my knowledge and skills would help me overcome these difficulties. I had to count on my internal and external strengths. Physically, I was in great shape, and mentally, I focused on being prepared for everything - from handling child abuse cases to investigating traffic accidents. Being mentally prepared gave me the confidence I needed to be better at my job. I developed my inner strength by being confident. I refused to listen to the people telling me I shouldn't be an officer; instead, I focused on who I was helping in our city and worked toward being the best police officer I could be.

I loved helping others, and early in my career, I realized I could make a difference both in the communities I served and in changing my department for the better. I learned how to work with children and help those who were victims of crime, scared and worried about what would happen to them. I learned how to help others help themselves. I decided to work very hard to strengthen our communities and help other young women who wanted to join the department but didn't think they could. I hoped to make their journey a bit less difficult than mine. I also believed that building a police department that looked like the communities we work for would make us a better police department, a department of men and women of all races and all genders.

My hard work paid off. When I first came to the department, there were fewer than 200 women; when I left 35 years later, there were more than 1,900 women officers. We filled every rank: officer, detective, sergeant, lieutenant, captain, commander, deputy, and assistant chief. The LAPD is led by the Chief of Police and three assistant chiefs. My dreams pushed me to study hard, and I was promoted through the ranks to become an Assistant Chief, an important and powerful position where I led many officers in the department.

Through all that hard work and accomplishments, I never lost my sense of adventure and love of science. I dove the Galapagos Islands, the Caribbean, and the south of Spain. In Africa, I went on a camera safari to Tanzania to experience the beautiful wildlife and fell in love with the people. They were so wonderful and kind, teaching us about the history of Africa, their culture, and the various animals that fill their lands.

I was fortunate to have close friends and family who supported my goals and always encouraged me to excel. My hard work and determination were significant factors in my success. Most importantly, sticking with my dreams and goals and not giving up despite challenges led me on this wonderful journey. I loved what I did so much that even though I am retired, I stay involved as a volunteer police officer so I can continue to help others. I travel all over the country, helping police departments learn from my experiences by speaking and teaching how to manage conflict, community policing, preventing violence, and leadership.

Please know how powerful each and every one of you is. Understand that there are people who believe in you. When times seem difficult or even impossible, recall the influential people in your life. They are with you in spirit, supporting the pursuit of your dreams, pushing you to do things others say are impossible, going places you never thought you could, and helping others in need or who just need a little guidance or a kind word. Most importantly, never forget that you are in charge of your destiny, and always DREAM BIG.

"Please know how powerful each and every one of you is. Understand that there are people who believe in you."

METEOROLOGIST-IN-CHARGE

FELECIA BOWSER

"As much as possible, surround yourself with people who believe in you, care for you, and want the best for you."

Since I was seven years old, I have wanted to become a meteorologist (someone who studies the weather). It started with my parents gifting me a weather book for my birthday. Reading about how clouds form and what happens when a tornado develops fascinated me. My parents nurtured my enthusiasm by taking me to the library to read more about the weather. They always encouraged me to learn as much as possible.

When I started high school, a science teacher recognized my passion and gave me videos that explained how hurricanes and tornadoes formed, giving me an inside look into these dangerous and destructive weather events. My grades in high school were good, but I struggled with physics, a crucial subject for an advanced degree in meteorology. I did not let that stop my desire to learn because teachers were there to help me (even after class). I took full advantage of that in high school and college. Teachers in college have preset days and times called "office hours" when they are available to answer students' questions and give advice on the subjects they teach. I spent more time than I could count in my physics and mathematics teachers' offices, reviewing what they had taught that day. I did not want to go through college barely understanding a subject; I wanted to master what I learned because I knew I would need it for my future career.

One day, when I was a freshman in college, a mathematics teacher asked me to stay after class. She noticed I was having difficulty in her class, but instead of offering advice and help, she berated me. She said I should drop out of her class because I would probably fail. She also said I should drop out of school to work at a fast-food restaurant because it would only get harder from here, and my future did not look so bright. I could not believe a teacher would say such things to a student. Did those words hurt my feelings? Of course. However, she lit a fire in me, so I set out to prove her wrong! Looking back on the situation, I see that maybe her goal was I needed to step up or give up! I stepped up my efforts and passed her class. Afterward, when I saw her around campus, especially on days when I felt tired, seeing her gave me an energy boost.

"Volunteer in the area of work you are considering to get a taste of what it would be like as a career."

In my first year of college, there were 300 students in the meteorology program. Four years later, only 30 of us received our diplomas by graduation day. I had proved her wrong! Whenever I wanted to give up, I thought of my teacher's words. I realized that I had the strength to be successful all along. It wasn't her words that pushed me; it was my drive and dedication. I pushed myself. I knew I could do it with hard work and willpower, and I did it!

Some people may tell you that math and science are hard and that maybe you should pick an "easier" subject to study. Do not let people who do not want to see you succeed or do not share your vision decide your fate. As much as possible, surround yourself with people who believe in you, care for you, and want the best for you. In addition to my family, I was fortunate to have a very tight-knit group of friends in college who supported one another. When someone in our group felt down, we would lift that person up with our words and actions. We never gave up on each other. Once, we were pulling an "all-nighter" (staying up all night studying for an exam). After the fourth hour of studying, we needed to take a break. As we got up to stretch and grab a coffee, we decided we needed a team mantra to get us through the night. We began to chant "We are going to make it!" from a popular song at the time. We made it through the night; the next day, we all passed with flying colors! That verse became our mantra to keep us motivated throughout the rest of our college years. Still, when I hear that song today, it puts a smile on my face.

I now work for one of the nation's most notable and famous meteorology organizations. I have successfully climbed the ranks and become a management team member. I give talks about meteorology to elementary and junior high school students. I speak about my organization, my position, and what it takes to become a meteorologist. Students are often shocked to see a female meteorologist. I tell them I am proof that young ladies can be anything they want to be, no matter the profession. Girls often come up to me for a hug and tell me that they, too, want to become meteorologists and that I am a role model. Performing these school talks is work to me, but it doesn't feel like it and is an honor. I am privileged to have a positive impact on my community by doing something that is part of my job, which is not only great but also fun. Do I ever think back to my old college teacher's words? Sometimes – usually when I walk into a fast-food restaurant.

CONSTRUCTION MATERIAL TESTING NATIONAL PROJECT MANAGER

KIMBERLY SMITH

"It's okay to change your mind and not follow a straight path in life, but steadfastly follow your dreams. Never lose sight of what is important to you, no matter the challenges and obstacles you face."

I always wanted to be a "mad scientist" when I grew up. To me, that meant someone who broke away from what we believed to be true and challenged assumptions about the world. It also meant someone who spent time inventing, experimenting, discovering, and getting their hands dirty. I love science. It's exciting to conduct experiments even when some of them fail, and it's fun when facts change because new data becomes available. The possibilities for learning and discovering are endless. I can learn about the ground we walk on, the air we breathe, and the sky and stars we see at night. This is what sparked my passion for STEM and why I chose to dedicate my life to science. I wanted to learn everything I could about the Earth, and studying geology quenched my thirst.

I was diagnosed with Attention Deficit Disorder (ADD) as a child and learned that school would always be harder for me compared to my friends. ADD makes it difficult for me to focus and learn. A few years later, I was also diagnosed with scoliosis, a curve in the spine that made my daily functions, like getting dressed in the morning, difficult. For two years in middle school, I had to wear a brace that stretched from my shoulders all the way down to my hips. Afterward, I was allowed to go without bracing, but I needed to work on proper posture and monitor the curve in my back. If my curve gets past a 55-degree angle, I will need surgery. I am at 51 degrees now and being watched for changes. A doctor once told me that because of my disabilities, I would never make it far in life; I wouldn't be able to achieve my dreams. It broke my heart. I felt so defeated. Eventually, it made me stronger-willed and even more determined to succeed than before. I was motivated to prove him wrong.

As I grew older, my interests came into focus, and my dream of being a mad scientist changed into wanting to become a geologist, a scientist who studies the Earth's structure, its history, and what affects it. I was excited to learn something new and specialized and become the first person on my father's side of the family to attend college. It did not seem possible at first because of the cost. I realized the money I saved was not enough. My academic talents were strong, and I was honored with scholarships by my high school. Still, it was not enough. To help, I worked three jobs in college to support myself. While money was a huge factor in where I went to college, my passion for science and desire for a fruitful learning environment pushed me to excel. Goals are achievable when you set your mind to them!

"Even when it seems that a new challenge arises every day, your attitude, problem-solving skills, and vision for the future will guide you, help determine your success, and allow you to grow."

I was attracted to geology because it's a field not many people know about. It has a small number of experts, but geologists are growing in demand. It allows me to work outside in nature and with my hands. I learn off the land and about the Earth's past by looking at the natural features around me. Geologists are taught that "The present is the key to the past." We study the development and formation of a land area by inspecting and documenting the landscape and what it's made from. You can only do that by being outdoors. As I learned more about the environment, I came to understand that the Earth needs protection. I promised myself that I would protect it in any way I could because, as citizens of this world, that's our job. With that mindset, I grabbed an opportunity to help when a conservation spokeswoman gave a talk at my university about a species under threat of extinction, the piping plover. After her talk, I found the courage to ask her what I could do to help. It changed my life forever.

I began working with the Friends of Recreation, Conservation, and Environmental Stewardship (FORCES) program to repair beaches along the coastline of Lake Ontario. In 2016 and 2017, extensive flooding along the lake shore washed away many precious nesting grounds for the plovers and flooded local homes. I was devastated when I saw the impact of the water. After the water levels went down, the coastline was walkable and visible again. We spent countless hours cleaning the beach and removing harmful garbage. Entire boat docks, other wooden scraps, clothes, shoes, toothbrushes, trash bags, and even car tires and bumpers were scattered across the beach. We spent three days placing fencing along the sand dunes to help the sand pile up and rebuild the land to provide greater protection to the birds' nesting grounds. The fencing also closed off some public beaches so the plovers could safely rebuild their nests and raise their chicks.

We planted single-stem dune grass plugs one by one in certain areas by the shore to slow the wearing away of the protective dunes. Staying off the high dunes is important because any footstep wears them away and can severely harm the ecosystem. The dune grass and other plant life growing on a beach prevent the wind from sweeping away the sand and make the dunes firmer when the water table rises. To protect this environment and the fragile creatures that live in it, we must always clean up after ourselves and ensure there is no garbage left behind. All dogs must be leashed because they can smell chick eggs, trample, and kill them. The color of shorebird eggs blends in with the sand and are easily walked on or damaged by mistake, which can lead to the death of those chicks. Always looking where you are walking on the beach is very important.

Almost immediately, the shoreline was cleaner, there was more healthy vegetation along the lake, and we observed the occasional piping plover landing in reserved areas but not yet nesting. However, most of the effects of our work won't be seen for a very long time. It will take a few years before the shore becomes completely garbage-free, assuming it doesn't flood again. It will take five to twenty years before the dunes build up high enough to provide protection. To further help the birds and the state park, I assessed the strength of the dunes and the health of the local plants compared to other areas that were eroding. With the help of state park officials, I used special equipment that took images under the surface of the beach to understand how the flooding and high-water levels changed the landscape. I compared pictures of the shoreline from the early 1970s to the recent ones I took to see how the land moved and shifted over time. I worked very hard and used that data to understand how piping plovers lived off the land and how the environment affected them.

I was the first person to do research in this area. I had to work long days and nights to find ways to research and study why the plovers' homes were being washed away and what could be done to stop it. I struggle with back pain when I work outside, and I have to remind myself to use good posture and stretching techniques to help with the pain. For me, this is a labor of love, so I stay focused on my work instead and the value it brings to the world. I have chosen to do whatever I can for others and the Earth.

This project taught me that although the beaches are fragile, there are ways we can protect them and the special creatures living there and many others like them. At the end of the project, my university honored me with the Innovation in STEM award for going above and beyond to create a research project that benefits the community and makes a lasting impact on the environment. My research earned me great friends and even greater memories.

In 2019, I moved from New York to Florida. Leaving my friends and extended family behind was one of the most challenging times of my life. I was heartbroken for a while and struggled to make friends and adjust to my new way of living. It took me ten months to find a job in my field. I sent out thousands of job applications, but networking, meeting people, and joining local organizations were the best ways to get an interview.

K. SMITH

I was hired as an Environmental Geologist and began learning the job requirements. Part of it includes typing reports and evaluating aerial photographs. With ADD, I struggle to concentrate and focus, so I find it best to work on small segments of a report at a time, shifting my focus every hour to different sections to be sure all tasks are completed on time.

During my job transition, I also had the opportunity to model women's clothes and enter the world of competitive beauty pageantry. I never thought modeling and pageantry were within my reach, but I took a chance that it would lead to personal growth. I ended up being crowned Ms. International World Italy, allowing me to create a platform to create more opportunities for women in STEM. Pageantry and modeling have helped me embrace my physical imperfections. I have learned that they make each of us unique and beautiful. It has taught me to stop imposing limits on myself, have faith in my abilities, and give myself time to grow. With growth comes prosperity.

Also, during this time, my mother was a nurse in a hospital during the COVID-19 pandemic, contracted the virus, and passed away. I was 23 years old and felt like my world crumbled around me. She was a foundational piece of my life. As a single parent, she gave me the tools to be a strong, independent woman. Losing her put many responsibilities on my shoulders that I was unprepared for. Dealing with her loss was difficult mentally and physically. I used the turmoil within me to drive a greater passion for my work. I set out to earn training certifications within the construction industry to gain more skills. My company recognized my efforts by allowing me to expand the company by opening in another location. I realized I love the technical parts of my work but also the development and growth side.

Furthermore, I used my ability to connect with others and build lasting relationships to get promoted to a national-level job requiring me to relocate to Atlanta, GA. This transition taught me that success is not just having a job I love. I became fulfilled in my career by understanding how to use my strengths to grow professionally and understand my limitations. For example, I provide customers who build buildings, apartment complexes, restaurants, and hospitals with information on what lies beneath the Earth's surface. This is important to land developers as it tells them how stable the ground is for building. This is especially true in areas where there are earthquakes, where the water table is close to the ground's surface, and where the strength and stability of the ground depend on how much water is in the area.

My ability to connect with others personally helps me build strong relationships, leading to more customers and repeat business. My problem-solving skills allow me to assist my co-workers on their technical projects. Geology has more career paths than I realized! These various opportunities and my love of continuing

K. SMITH

education drove me to begin working towards a master's degree in construction management. Being a woman in the construction industry is very challenging. It can be hard to find your voice and be respected, but taking an idea for a project and seeing it come to life through land development is a captivating experience and worth the effort!

Facing challenges is a part of life, inspiring and encouraging me daily to excel. Even when new challenges continuously arise, your attitude, problem-solving skills, and vision for the future will guide you, allow you to grow, and determine your success. Follow your dreams and pursue your passions, even if personal and worldly obstacles seem in your way.

"I have chosen to do whatever I can for others and for the Earth."

STEM TEACHER AEROSPACE EDUCATION MEMBER CHILDREN'S BOOK AUTHOR

SELINA WEBB

"People can and will help you realize your full potential, but you must be bold and try. Do not be afraid; even if at first you do not succeed, at least you know fear did not get in your way. Ultimately, you must put forth the effort, stay disciplined, and achieve your goals."

I was raised in Texas by wonderful parents who valued education. My parents did not have the chance to go to college, but they wanted to provide that opportunity to my four sisters and me, believing it would be the key to our future success. We did it! All five of us made it to college and graduated. My parents supported us financially and encouraged our efforts. They taught me to take every opportunity because it may not come again. If I believed I could achieve it, I should try to do it. Because of that attitude, I juggled many activities in high school. I was part of the United States Academic Decathlon (a yearly high-school academic competition), the tennis team, and the Spanish club. I took Advanced Placement courses (higher-level classes to get college credit). I enjoyed my friends, playing sports, and planning my future. Being part of a sports team gave me a sense of belonging. It also helped me become more disciplined, which led to other academic opportunities. I had to stay focused and maintain a high-grade Point Average to get into college. Many teachers and coaches helped and encouraged me along the way. They were great role models and wanted me to succeed. They could see how driven I was and that I had a strong work ethic. People can and will help you realize your full potential, but you must be bold and try. Do not be afraid; even if you do not succeed at first, at least you know fear did not get in your way. Ultimately, you must put forth the effort and stay disciplined to achieve your goals.

Goal setting is similar to playing a sport. The more you practice, the better you become. Keep your focus by staying away from bad influences (like drugs and alcohol). Never was this clearer for me than in high school. There were two paths I could take: I could have viewed the successful students as smarter than me and given up, or I could study, work hard, ask for help, and know that I, too, can be successful. I took the second path. I didn't let other people place limits on me, so don't let other people limit you. You can determine what skills and abilities to develop.

"Goal setting is similar to playing a sport. The more you practice, the better you become."

To stay focused and confident, there will be times when you have to ignore the negative things people say to you. A male student once said, "I can't believe you're ahead of me [in rank]; you don't even take advanced math." I felt surprised and angry that he thought he was better than me just because he was a boy. At the same time, I was proud of myself and what I had accomplished in math and at school, so those negative feelings did not linger. Others may judge your academic abilities based on your gender. Learning to use negative comments to motivate yourself to succeed is important. You have to rise above them.

My confidence and positive outlook grew from my belief that a higher power was looking out for me. Sometimes, I failed, but I kept going because I saw failure as a problem to solve, as an opportunity to gather information and learn. People often came into my life and encouraged me to surpass my hopes and dreams. For example, my boss recommended me to a program to earn my master's degree. His boss had done the same thing for him, and he wanted to return the favor by helping me. With his support, I completed my degree while working as an academic coordinator at the same time. In that role, I helped students who had applied but had yet to be accepted to the university or were thinking of applying, wanted to understand available majors of study, scholarship information, or how to enroll in college if they couldn't afford it. Sometimes, they were the first in their family to do so. Sometimes, they didn't know how to apply or lacked the confidence to get started. It was very rewarding to help young adults like myself not long ago take such a critical step toward achieving their goals.

Give back when you can, and you will be rewarded beyond measure. I felt blessed knowing these students took the first step to a brighter future and towards fulfilling their dreams with a little help from me. My experiences were lessons preparing me for my purpose in life: to help others achieve their academic goals by being a teacher. Be on the lookout for your lessons, too.

If you show up to a classroom or job site where you are the only girl in the room, don't be discouraged or scared; consider yourself blessed. You are unique, you are capable, and you can succeed. It is your turn now. Accept the challenge to exceed your expectations and follow your dreams!

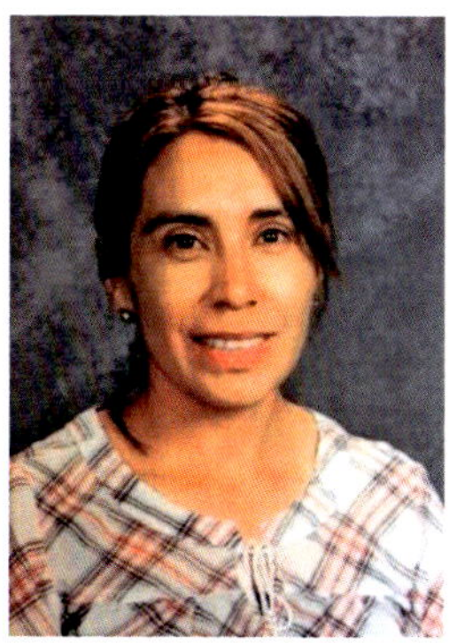

PRESIDENT AND CEO

LAURIE HALLORAN

"My hometown was too small and boring for my craving to see the world and make a difference."

When I was in middle school, human biology fascinated me. There were pictures in textbooks showing everything that went on under the skin. I was enthralled and wanted to learn the "Why?" of everything. I also loved baking, so I enjoyed learning about chemistry. Chemical reactions were predictable and measurable. There was a recipe and an outcome. It was a wonderful escape from my crazy household. I had a chemistry set that I kept high on a shelf so my younger brothers and sisters couldn't mess with it. It was mine and not for anyone else but me! Having something that was just for me was rare in my busy family, where everything had to be given equally. Since my things would often get broken or lost, I felt special and grown up whenever I had something all my own.

We played hospital quite often in my parents' basement. When the little kids pretended to be patients, I could always get them to take a nap. This was usually quite difficult with so much happening in our house. It made me feel triumphant! Plus, it meant it would be quiet for a while and no fighting! My mom thought I should be a nurse because I liked science and could always get the kids to fall asleep because I was so good at caring for them. I disagreed and really thought it just meant I was a good babysitter! Back then, most girls became nurses, teachers, or secretaries. My dad thought I should be a doctor because he said that was a more important job, but I thought doctors had to go to school for way too long. Like my parents, I expected to get married and have babies right after college. So, I went to Russell Sage College in my hometown to be a nurse.

I suggest moving away from home for college and staying in the dormitories if you have the means to do so. We did not, so while in college, I dashed home to make dinner for my siblings. This was not my idea of college life. I moved into the dorms on campus for the last two years of college, but most of the other students' friendships had already been formed by then. I had little opportunity to make friends, coming into the dorms two years late. Also, it was not much fun because I had to get up at 5 a.m. to get to the hospital for my student nursing.

The first time I cared for patients, I realized I really didn't like doing the work. It did not excite me like I thought it would, and I hated being with so many old and sick people. It was physically hard work without any challenge to my brain. I did not feel like I was making a meaningful difference at the end of a patient's life; I felt guilty for not liking it more. I was halfway through my nurse's training, and since I had no other ideas, I kept going. My first job as a nurse after graduation was caring for small children. It was good for me because I could see that I was making their lives better, but I was doing the same thing every day. I was still living at home even though I was 25 years old, and my life did not feel like it was going anywhere. I was very unhappy. So, I switched jobs to a bigger hospital. I moved to an apartment with my girlfriends and worked in the Pediatric Intensive Care Unit (PICU). It was challenging because the patients were very sick. There was much more for me to learn, and I could see the effect of my work when a child came out from a coma and opened their eyes. I worked there for three years. I still wanted something new and fresh, a bigger, more exciting city, especially farther away from home, and a job that was not nursing.

I decided to move to Boston because I thought it would allow me to find a new career. I had no idea what "it" was and was very nervous but excited. My hometown was too small and boring for my craving to see the world and make a difference. Moving to Boston was a big deal for my parents because they liked keeping everyone nearby. I needed to get away from my family and live someplace new. I needed to be independent. Boston was a natural choice for me. When I was in 8th grade, my school took a field trip to the New England Aquarium, and as soon as I got off the bus, I felt like I was "home." It was much bigger than my hometown of Troy, New York, and on the Atlantic Ocean with beautiful waves and beaches. Adding to the excitement in my young mind, Europe was just on the other side of that ocean! It didn't have to be a place I only visited in my imagination or through books. The world was out there for me to explore! There was so much to do and see. It was so beautiful that I just wanted to learn everything I could about it. I loved that Boston's history started with the birth of our nation. I fell in love with the uneven streets as we walked the Freedom Trail. Historical events I learned about in school took place in the buildings right before me! I kept saying, "It's not too far away from home, so you can get back quickly if you need to."

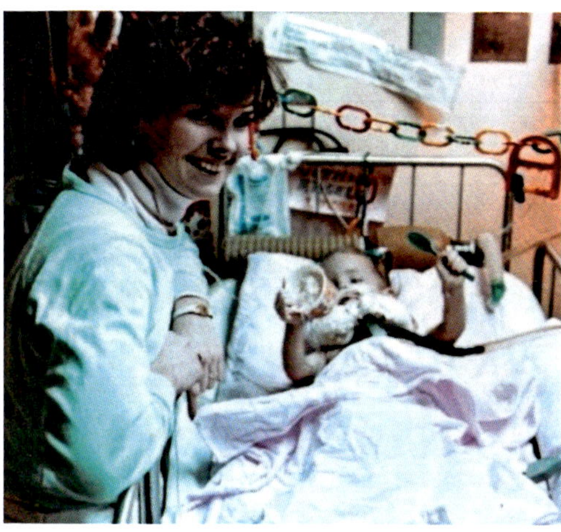

I moved in with two other girls I knew before, so it was a little less scary than moving to a new city alone. Once in a bigger city, I thought I would find a job that still allowed me to help patients, but it would be "different." I wasn't sure what that was yet, but I had to discover it for myself. My parents were sad, but they were supportive and helped me move.

My first job was at Boston Children's Hospital as an ICU nurse; it was always meant to be a short-term job. I mainly worked with babies born with heart problems. Their parents flew with them to Boston from all over the world to have surgery and hopefully lead a healthy life with a healthy heart. Sadly, many died from complications after surgery. Too many. Some of them right before my eyes. It was worse than my job before moving because if their hearts didn't work, I could do nothing to save them. It was very difficult work and really sad. Since I did not leave home to be in another position where I felt helpless, I planned to find a job where I had the power to make a difference. I started to look for a new job and found one quickly.

My next job was working for the city of Boston on a research project called "Lead-Free Kids," funded by the US Environmental Protection Agency (EPA). I was going to be a "research nurse." Whatever that meant – it was new to me, and at least it was not in a hospital! Boston had a big problem with lead poisoning in some groups of children, especially those in poor neighborhoods. When children are exposed to lead, it can cause lifelong health problems and brain damage. This project tested the theory that children were poisoned by playing outside where the paint on their houses was flaking off, contaminating the soil around them. We were going to see if replacing the soil would decrease the lead levels in their blood. Working on the project brought me to some very poor areas, and there were issues with gangs, drugs, and gun violence, sometimes happening right in front of me. I was the only white nurse on the team and had never been to these parts of the city before.

We went door-to-door to get children in the study. Many of the neighborhoods we visited had entire city blocks of empty lots and abandoned houses. I saw gunfights and homeless people. I saw my black coworkers treated differently when we went to cash our paychecks at the bank. It was quite an eye-opening experience. I learned more about racism and poverty in 10 months than in the 27 years before. I also learned about research, calculating the proper number of people needed for a good study (sample sizes), and finding the right people for a study. We started to hear rumors that the EPA would stop the lead study. I began worrying about what I would do if that happened, so I started to look for a new job again.

I saw an advertisement in the Help Wanted section of the newspaper that read: "Research Nurse 50% travel," I said to myself, "Wow! There might be a company car if I need to drive around a lot, and I really need a new car!" I called and found out it was airline travel throughout the US to different hospitals and that I would be working on studies of new medicines as a Clinical Research Associate (CRA). I love to travel, so excitedly I interviewed and was hired. The job required me to fly to hospitals and taking part in a study. Once there, I checked and confirmed the data collected to prove that the medicine in the study worked and was safe. I had no idea this was how new medicines got to patients! This was an entirely new career where I could use what I had learned as a nurse to help patients in need. But I had a lot to learn about clinical research.

"Listening is much more important than talking. You will learn a lot if you let other people speak."

In the US, the Food and Drug Administration (FDA) has rules about how companies design and perform research studies on any new product to treat or cure diseases. Those rules require research studies to be completed in laboratories first (without contact with patients), followed by studies in animals, and then, only when the potential benefit is greater than the risk, does the FDA allow people to be in a study. The stage where people are studied is called clinical research. In this stage, researchers create a clinical study protocol. The protocol is like a detailed set of instructions for playing a game with very specific rules about who can play, how to play, and how to collect points to show you followed the rules. Then, they find expert and highly ethical doctors from all over the world who have patients with a specific disease, and those patients agree (consent) to be in the study. The clinical study usually compares one set of patients who get the investigational (unproven) drug to an equal number of patients who are given the current treatment (or sometimes nothing, called a placebo). During clinical studies, doctors ensure they aren't taking too much risk and hurting patients. They must document everything according to very high standards, called Good Clinical Practice (GCP). The doctors and their teams collect data on how the study patients' health improves, as well as any negative side effects. Thousands of patients might be studied, and someone must monitor what is happening to ensure the doctors did everything right and collect the data to submit to the FDA. That is the job of a CRA.

I was excited that I had found the job I was looking for! I was helping patients receive better treatments to make them feel better or even be cured. Finally, I was making a difference by helping bring new cures to patients in a way I never could before. And on top of that, I was seeing the world while I was going to work!

After traveling to dozens of cities, I spent almost a year in California studying patients who were suddenly getting a rare form of pneumonia and dying. We didn't realize it yet, but they were infected with Human Immunodeficiency Virus (HIV) and were dying of Acquired Immune Deficiency Syndrome (AIDS). After months of flying back and forth, California had become almost like home. I especially loved San Francisco. My co-workers and I started spending weekends in beautiful places along the coast rather than flying back to Boston. We worked really hard, but seeing all of the beauty along the California coast each weekend was wonderful.

It was extremely rewarding because the drug was approved rapidly by the FDA based directly on our work, and it was given to patients to help them recover from pneumonia. I continued to work on HIV drugs for years afterward, and I was thrilled when it became a treatable disease instead of a death sentence.

For ten years, I worked at the same company on new medicines. Many were for diseases without any other treatment, like certain types of cancer. I kept learning, and every day was different. The company had grown, and they wanted me to travel to Europe every month. When I was young and single, the travel was a blast, but I had gotten married and was going to have a baby. I wanted to be in Boston with my family, so I started my own company. I didn't really have a plan, and when my son was born, I realized I needed daycare and needed work to pay for it! I learned the hard way that it's important to have a plan before making big decisions like starting a business.

So, while developing my plans, I started working at a small local biotechnology company, making an entirely new type of cancer drug. It was even more interesting because I got to work directly with the FDA. I learned a lot about how to manage a program to develop a new drug and how biotech companies work. We only had one drug we were developing, and while I was there, a large pharmaceutical company bought it from us. I had worked hard on advancing the drug and was sorry to see it go because I was very connected to the project. At the same time, I was going to have my second baby, and my company didn't have paid maternity leave. I could not stay home with my two babies and have enough money to pay the bills. I needed to look for a new job again. This time, I was looking for a company with more than one product so I could stick around longer.

When my second baby was just three weeks old, I went to work for a different company to afford to send my two-year-old to daycare and pay the mortgage on my house. Even though it was a bigger company than the last one before I arrived, it was mostly made up of business people and scientists. There were no experienced clinical researchers. After I saw what had been ignored or overlooked, I realized that even if their drug worked, they hadn't done the studies by the rules and regulations. FDA would probably reject their data. I began to highlight the problems and ways to fix them while they still had a chance. It mattered to me that the rules were followed because I was the responsible person. I wouldn't sign my name and send data for approval to the FDA if there were ethical issues or the rules were "bent." After almost two years, they did not support my recommendations to do the studies according to GCP, so it was time for me to go.

I was tired of other people not knowing, not caring, or ignoring the rules set in place. I suspected other small companies had similar problems. Companies that didn't have the right experts and probably needed them very early in the development of a product, just like my last two companies. I thought I could make a real difference to small companies by offering a team of experts to advise on how the FDA expects to see things done. If company founders learned how to make good decisions early on, we could help bring new medicines to patients. We would always do what was right for the patient; the company employees would be more successful and make money for their investors. Suddenly, I was the President of a company. I remember as a kid, when we weren't playing hospital, we were starting up a neighborhood club, lemonade stand, or finding and reselling golf balls at the local public park; I always wanted to be the one in charge. I got what I asked for! I used to think that because I wanted to get married and have babies, I couldn't be a doctor because that was "too big" of a job. But I didn't have my first baby until I was the President of my own company. I realized I could have a baby and a BIG job!

"While it's tough to say you don't know how to do something, it's the best way to learn. Ask for help when you don't know the answer and ask for feedback to become a better person and leader."

I did not go to school and learn how to create a company, but I knew enough to get started. Business school can't teach you what to do when you are leading a company and have a problem. I had to keep learning from others, reading about good leadership and management skills, and becoming the best at my job. I also spent time with other business leaders to learn from their experiences. In life, you never finish learning. But to be a great leader, you must figure out what you love to do and do well and what you don't. Just like George Washington, who said he needed a cabinet because he wasn't good at everything, once you figure out your weaknesses, you can partner with people who are great at doing what you can't! No one is an expert at everything. Knowing your strengths and understanding and admitting your weaknesses will make you a good leader. While it's tough to say you don't know how to do something, it's the best way to learn. Ask for help when you don't know the answer and ask for feedback to become a better person and leader. Surround yourself with people who are good at things that you are not, and as a team, you are much better than any one person could ever be. Most problems are caused by people who don't lead well or can't admit they don't know something.

Over 20 years ago, I started my company in an unfinished half bathroom that we called the "Closet of Doom." We moved locations many times as the company grew, and now we have over 100 employees and work in two beautiful offices in downtown Boston and La Jolla, California! As President, one of my most important jobs is ensuring my employees feel important, cared for, and are able to work on exciting projects. My other top jobs are to make sure the company has the leadership it needs now and for our future and to make sure we are prepared for the needs of our clients. I meet with every new employee after they start. I ask them how they got here, what they love, and what they are passionate about.

I explain that I sincerely want their feedback, especially if they have ideas for improving things. I answer their questions about me honestly and ask what I can do to make their career special. Most people want to know the long-term plan for the company. At this point, I could sell the company, retire, and live happily ever after, but that would be selfish. We are working on becoming an employee-owned business to unify ourselves with common goals and face them and the future together.

As a team, we do fun things together as much as possible: dinner out, playing games and having contests like scavenger hunts. We also support charity events to support that the employees select as important and meaningful. We work hard to make everyone feel included, no matter their background, their interests, or who they love.

Every Friday, we have a company-wide call for "Weekly Roundup and Praise" (WRAP), where employees submit good news, give thanks and appreciation to each other, and discuss how much they enjoyed helping on projects. Our work is difficult, so it takes dedicated and caring employees who aren't afraid to speak their minds. The single most important thing I can do is back them up when they give advice; even if it isn't perfect, it's their best thinking.

After many years of growing and learning, I still feel like I have many new areas to explore and ways we can make a difference, so I am never bored. I am good at listening to other people's input, learning from their ideas and experiences, and putting myself "in their shoes" or empathizing with them. I can see what is important to them and what makes them nervous. It helps me to understand what they want. I also learned that I need to speak up. It's hard to be confident all the time, but believe in yourself, listen to your "inner voice," and don't be afraid to say what needs to be said. Always be on the hunt for good role models, observe others to learn what they do well, and borrow what you can from them, but make it your own. Listening is much more important than talking. You will learn a lot if you let other people speak. And the most important lesson is that you are never done learning! Where you start in school is not always where you end up. Don't think that just because you went to college or got a job doing one thing, you must stay there if you have new dreams or ideas. Trying and failing at something new is much better than sticking with something that does not make you happy.

Now, my kids are grown up and need different kinds of help from me about their lives, loves, and careers. I am pleased they still seek me out for advice or to have fun together. To me, that is my biggest success: to have raised happy, healthy, and fulfilled young men who feel like I am a good mom and that I also make the world a better place in my own small way.

Girls are often told what their path should be by parents, teachers, or friends. I would say never to be afraid to choose a different path if the first one doesn't feel right. Your life isn't a ladder to climb; it's a long and winding hike. Some people want to climb a mountain, and that's fine, but you can also be happy walking on the beach at a slower and simpler pace or hiking in a desert (still challenging, but maybe off the beaten path). If you are looking for help, many people can give it to you, but you have to figure out what you need to be true to yourself. Learning is a lifelong journey, and you have just begun. If you can't wait to stop learning, you'll be stuck where you are and won't grow in life.

L. HALLORAN

GIVING BACK

One of my new projects is working with a group of companies in Boston to help college students who are not financially well-off get into clinical research through internships. It's called Project Onramp. I also started a scholarship program in honor of a wonderful female friend, a pharmaceutical company employee, who died soon after she had a massive stroke. The scholarship is for women pursuing leadership education to lead life science companies. I want to give back by sharing what I have learned with newer companies as a board member (advisor).

SENIOR SUPPLIER QUALITY ENGINEER

NIKITA ANGANE

"I am very proud of what I have accomplished in the US, leaving behind the comfort of my friends and family."

What if I told you that I am an engineer and hate math? I even flunked math in fourth grade. If my school hadn't allowed me to retake the test, I would not have been promoted to the fifth grade. This particular incident made me fear math. My parents hired tutors to help me; although they helped me to achieve good enough grades to pass each test, I never got comfortable working on math. I avoided it and cried when I was forced to complete my homework. The hatred and fear kept piling up. I was waiting to finish high school so I could leave math behind me forever, or so I thought. I decided to become a medical doctor to avoid the bitter reality of mathematics. This decision was easy because I genuinely enjoyed biology, an important part of training to be a doctor. Well, as they say, one must never say never.

Unfortunately, I didn't do well in my medical entrance exams. I wanted to stay and work in medicine, so I planned to retake the exams the following year. However, I had a hard time deciding what career path to take. My friends and family encouraged me to take up engineering since my test scores were good enough for engineering school. I opened a brochure that described a degree in Biomedical Engineering. I was curious to learn about a program that combined the world of medicine (human anatomy, physiology) with engineering (mechanics, electronics, calculus). My mind was made up. Not long after that, I found myself in an anatomy laboratory holding human hearts and brains, connecting wires in my electrical lab, and solving calculus problems in my math class. I hadn't realized before how this blend of topics and studies could lead to inventions in medical technology. I discovered what I wanted to do for a career. I was elated to be in the medical world and thrilled to step out of my comfort zone and excel in engineering. In the senior year of my undergraduate studies, I was part of a team that built a system for early detection of nervous system problems caused by diabetes. I am so very proud of this project because, for the first time, I used all of my learnings to help improve people's lives. Deep inside, engineering was calling to me when I was chasing medicine.

I pursued a master's degree in engineering from a US university to "spread my wings" and be independent. I was thrilled by the idea of taking on a new adventure – becoming comfortable with a new culture, meeting new people of diverse backgrounds, building a sense of self away from my family and childhood home, and exploring what I could do to improve my skills in my field. It is not just the skills I would gain; such experiences shape you in unthinkable ways. I began a new life in a foreign country with new types of people. I took tremendous courage to step out of my parents' love nest and move thousands of miles from my birthplace in India. I was determined to do what I decided and would not back down. I ended up getting a master's degree in a field that I never even thought I wanted to study; I didn't know that field existed. Of course, life as an immigrant in a new country wasn't easy. You must change yourself to blend into your new world while keeping your identity. It was like taking a sharp turn. Everything changed overnight, from the weather and food to how I talked and spent my free time.

"Before starting my new life, my weekends were all about lazing around, but now my weekends are filled with a new sense of responsibility for my future."

Though it truly was a fun learning experience. Living abroad gave me a new identity. Suddenly, I could do certain things effortlessly, like completing a 10-page report on time (back to this greater sense of responsibility). On the other hand, there were some things I couldn't do despite putting all my energy into it, like driving a car. No joke, it took me months to gather the courage to drive on the freeway. That feeling of independence that comes from doing everything on your own, whether it's being financially stable (having enough money) and being able to buy what you need and want or being self-sufficient - like cooking your own food, driving yourself places, taking care of yourself when you are sick and much more. I couldn't cook before I made the move to the US. I was completely dependent on my mom to cook for me. But now, trust me - I can cook amazingly well. There is always a first time for everything; unless you try to do it on your own, you will never know if you can do it. There are people to help you on this journey, but you need to learn when to ask for help. You don't want to be the person who bothers everyone all the time with small problems or requests, even if they are your friends.

N. ANGANE

Along with my newfound sense of responsibility, my competitive nature drives me to do things better. It could be scoring good grades in school or acing my work in the office. There is competition everywhere in today's world. It can be your true friend by keeping you focused and motivated, so long as you take it positively.

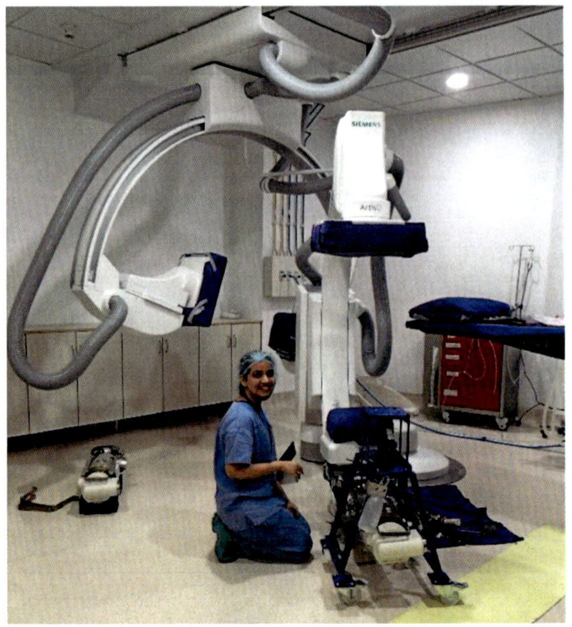

"There is always a first time for everything; unless you try to do it on your own, you will never know if you can do it."

While pursuing my master's degree, I worked part-time as a quality engineer for a medical device company; managing two lives – one as a student and one as a professional – was difficult. This was especially true when managing my time. There were sleepless nights and many cups of coffee juggling my work, school assignments, and tests. However, it felt fantastic to learn a new concept in the classroom one day and then apply that knowledge the very next day at work. It was exciting and satisfying. It was a fun ride; I was fueling my desire for more knowledge, on the one hand, and improving my skills and expertise in the professional and business aspects, on the other hand. This point in my life was precisely when I discovered how I wanted the rest of my life to look like - marching toward my dream of being known and respected as a quality expert for medical devices.

My STEM story shows you that hard work pays off. I am now a Lead Quality Engineer for medical devices. Every day, I ensure that only quality medical devices reach patients and improve their quality of life. I also lead a team of engineers that helps medical device makers bring new products to stores and hospitals. Highly skilled people, including medical doctors, look up to me as an expert and seek my advice on the safety and quality of their products. I did not become a medical doctor, but I am incredibly proud of how I reached a leadership role in a field I love.

DISTINGUISHED PROFESSOR OF EARTH SCIENCES

LAURA GUERTIN

"...I've always been a very stubborn person and never shied away from pursuing things I was really interested in. I believed in myself and my passion for learning and conducting science experiments."

I started my life uninterested in anything to do with science. This wasn't surprising to me, as I had no scientists in my family and had never met one. I surely never saw a woman as a scientist in cartoons or books, so I never imagined I could become one. Then, I had an amazing female high school chemistry teacher who believed in me because I could quickly and accurately solve problems and complete laboratory experiments. She told me I could do whatever I wanted in science and that I was good at it, too. Thanks to her encouragement and confidence in my abilities, I went to college for a science degree. My journey beyond high school was not a smooth one. I hit a few road bumps and detours along the way, including more than one college professor telling me, "You're not smart enough" and "You won't be successful because you are going to drop out of school, get married and have kids before you finish your degree." But I have always been very stubborn and never shied away from following things I was interested in. I believed in myself and my passion for learning and conducting science experiments.

"Don't let others dictate your choice of major [in college] or career. Seek advice and ask questions of others. Take time to explore a variety of fields (science, art, science communication, science, law)."

With my excitement about learning science and a bit of stubbornness thrown in to get me through the courses, I earned an undergraduate degree in geology. I went on to complete a PhD in marine geology and geophysics. I have been a university professor of earth sciences for over twenty years. I have earned local and national awards for my excellence in teaching. This has been incredibly rewarding since I care deeply about inspiring students to learn earth science, and having my teaching methods recognized by others outside my university shows that all of my hard work was worthwhile. I have received research grants from the National Science Foundation (NSF) and the Environmental Protection Agency (EPA) to explore how new and emerging technologies can be used to help students learn earth science in a university classroom. I am out in the community, speaking at museums and even in front of City Hall in Philadelphia about my experiences and passion for science to people who think they do not like science and it's not important. By all measures, I am a successful female scientist with an established career. I balance the "science" side of my life with fun activities. It's good to get away from scientific research occasionally, give my mind a break, and explore fresh ideas. Sometimes, stepping away energizes me and makes me more productive when I return to work. For example, I love basketball and attending home games of the Philadelphia 76ers - even though I grew up a Boston Celtics fan! I enjoy hiking and traveling to national parks, from Everglades National Park in Florida to Glacier Bay National Park in Alaska. I also crochet and quilt. My projects with yarn and fabric are mainly focused on creating items I can donate to children's charities or give as gifts. But I recently saw a way to merge my creative hobbies with my analytical, scientific self. You may have heard of STEAM, where STEM fields are connected with art.

"Although I always had a fascination with dinosaurs and the ocean...I never connected them with science or that a science career could allow me to study them full time."

Many kinds of artwork are made from scientific data and stories, from a crocheted coral reef to a musical performance of temperature data on the cello. One day, while exploring online, I saw a blanket someone crocheted patterned after a year's worth of temperature data! The blanket had 365 rows, and each row was color-coded to the highest temperature recorded in a specific location for that day of the year. I had never seen anything like this before. Since I crochet, I was drawn to the idea of connecting my creative hobby with my scientific background. Immediately, ideas started swirling through my mind–what other data could be represented with yarn? What else could be made besides a blanket? The possibilities were so exciting to imagine, and the options of what type of data to showcase with yarn seemed endless. I immediately bought a range of colored yarns, looked up different kinds of data online (temperature, precipitation, snowfall, etc.), and started crocheting. I made a scarf with temperature data from where I was born (Springfield, MA) and showed the temperature record for the first year of my life. I looked up the temperature data for where I live now for the current year, fifty years ago, and one hundred years ago, and crocheted three scarves. I brought these three crocheted temperature records to my classroom and showed them to my students for comparison.

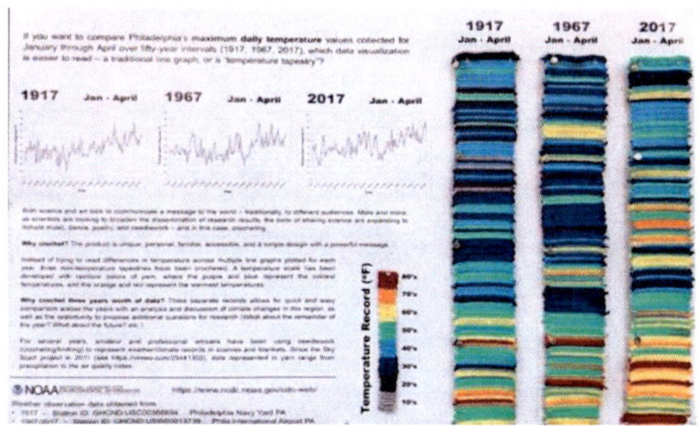

They instantly started shouting out observations about the different colors (temperatures) during different times of the year, how the colors differed from year to year, and so forth. I was so excited to see that my students could read temperature data so easily and with such interest instead of just looking at line graphs and figures of data in textbooks. I decided to share my crocheted temperature data at a scientific conference. I was really nervous about bringing an art project to a scientific meeting. What would the other scientists think? Would they want to see this intersection of science and art and how crocheting could communicate data? Or would they think creative arts do not belong at a science conference? Would they laugh at me and think I was too "girly" because I crochet? Fortunately, all of my fears were for nothing. The scientists were excited to learn about my work, discuss my students' reactions, and share ideas of other ways data could be explained through art. What a relief! Their reaction at the conference encouraged me to explore how else I could combine science and art.

At another conference, I was part of a group trying to describe how the biosphere (the regions of the earth occupied by living things) in coastal Louisiana adapted and became stronger over time. We were trying to find ways to share "coastal optimism"–stories of how humans and animals cope with a quickly changing physical environment due to habitat loss, saltwater intrusion, increasing storm frequency, etc. I immediately wondered if I could use crocheting to tell stories of how scientists and residents stabilized their eroding and sinking shoreline. I struggled to find ways to use yarn to tell these stories and tried to think differently about the challenge. My other artistic hobby, quilting, became my next communication tool. I created a collection of quilts called "Stitching Hope for the Coast." Each quilt tells a different story of how scientists and residents are successfully working with the ever-changing coastal environment of Louisiana. For example, parishes in Louisiana use discarded Christmas trees to make offshore barriers that lessen wave energy from wearing away the coast. The Pointe-Au-Chien Indian Tribe plans to construct greenhouses at a higher height to grow food and medicinal plants, as their traditional gardens struggle to grow when saltwater gets into their fields during storms. Even the TABASCO® factory, headquartered in coastal Louisiana, is slowly restoring the marshes by planting native grasses and working with residents and community organizations to bring back the local wildlife.

Like my crocheted temperature data, everyone from scientists to local crafters is excited to see science stories in quilts. I now have quilts detailing ice loss in the Arctic and the impact of warming temperatures on Alaska's Iditarod Sled Dog Race, and entire quilt collections on global warming solutions, the Endangered Species Act, and deep-sea ocean exploration. I found comfort and acceptance in being a scientist who shows her creative side through the arts. I discovered the joy of engaging with STEAM later in life. I was incredibly nervous about sharing the connections I saw between science and art. I had convinced myself that people would view me as less of a scientist if I admitted that I could do art and enjoyed it. Fortunately, no one has questioned my scientific background or accomplishments. I encourage you to explore how science can connect with other areas like art, music, law, or policy…You name it! My unique way of looking at things allowed me to reach new audiences and inspire those around me to think differently about what it means to be a scientist.

Don't just think outside the box when approaching a scientific challenge; embrace a creative approach to any challenge. You may be surprised at the innovations that come from the intersections of subjects you never thought of before. Don't be afraid to express your creativity; believe in yourself and your abilities. No one has the same ones as you do!

CAPTAIN (RETIRED)

MICHELE FINN

Science and technology have been the focus of my life for as long as I can remember. I was fortunate to have teachers in high school who saw my aptitude, a father who refused to let me pursue any subject matter that was easy for me, and a desire to explore the world. When faced with decisions about where to go to college, what to study, and what I wanted to do with the rest of my life, I had no idea what to do. No one told me that you could adapt and change course throughout your life and that those early decisions were not final. I knew I wanted to pursue science and live by a body of water. So, I confidently stated that I wanted to study marine biology at Texas A&M at Galveston, but I really was not so confident. While I was there, I took full advantage of my time on the Gulf Coast of the United States by being part of the sailing team.

I took jobs that let me have fun and learn more about my field. At the same time, I made strong relationships with professors and students. I found out later that this is a very important skill. After three and a half years, I was preparing for graduation with a B+ grade point average. Once again, I was faced with decisions about what to do next. And once again, I had no idea how I would make them. I had two criteria for my first job out of college: 1) I wanted to travel and continue to learn, and 2) I needed to start making decent money.

I interviewed for every job I could find, especially those from the company recruiters who came to my college. Then, one day, a recruiter changed my life.

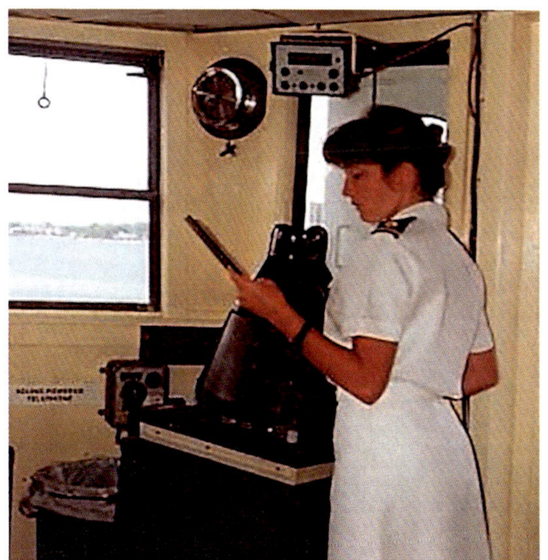

I learned about the National Oceanic and Atmospheric Administration's Commissioned Officer Corps (NOAA Corps). Wait! What is THAT? The NOAA Corps is one of the nation's seven uniformed services that traces its roots to the former US Coast and Geodetic Survey, dating back to 1807 and President Thomas Jefferson. I wanted to join the select cadre of professionals trained in engineering, earth sciences, oceanography, meteorology, fisheries science, and other related fields. Here is what really interested me: Corps officers operate ships, fly aircraft, manage research projects, and run scuba diving missions. Drive ships! Fly airplanes! Scuba Diving! I was SO in!

I applied, but I was not selected. I was designated as a backup, a second choice; I was disappointed. Still, I did everything they asked me to so I would be ready if they called upon me. At the same time, I had to finish my last semester of school and still find a job. I kept in touch with the NOAA recruiters, almost begging them to put me at the top of the second-choice list. Graduation came and went, and I was still the second choice. I was hired to start a fun short-term job, and I packed my car up to head to it when I got the call. If I could be ready to go to basic training in less than two weeks, I had my chance to be a NOAA Corps officer. I was so happy and so ready!

Before I knew it, I was sitting in a room with twenty-three other officer candidates, ready to learn how to drive ships, support scientific researchers, and be an officer. Five-and-a-half months later, I was assigned to a 133-foot research vessel as a deck officer. For two years, I worked with the very best group of fellow officers, crew members, and researchers, studying every bay, estuary, and mile of coastline from the tip of Maine to the border of Texas. My commanding officer taught me to operate the ship expertly in very tight spaces, with many other boats and ships around and gear hanging over the side. I also became an NOAA working diver and then a dive master. My performance on this assignment helped me become known as dependable, skilled, and determined. After two years, thanks to the advice of my senior officers, I was selected to fill a scientific operations position in Hawaii.

In that role, I supported groups of scientists studying endangered species, specifically the Hawaiian monk seal and sea turtles. I traveled to all the islands and atolls in the Northwestern Hawaiian Islands. We commanded ships, small boats, and aircraft, and we dove and camped in remote areas. I learned about marine mammals and sea turtle biology. I also became really good at operating and maintaining long-range communications equipment, solar panel systems, gas generators, small boat engines, and various field sampling equipment. At the same time, I took advantage of living close to the University of Hawaii and my relationships with professors to get a master of science degree in Zoology.

So, how could it get any better than a cool coastal ship and an extended tour in Hawaii? Flight school! The NOAA Corps sent me to flight school. I was selected because I took the initiative to obtain an advanced degree while working full-time (and with overtime) and because they needed competent women to join the program. I was not the first female aviator for NOAA, but there had not been many before me. I learned to fly to help scientists conduct airborne research. For example, I flew a De Havilland Twin Otter for marine mammal and Bluefin Tuna population surveys, air pollution studies, coastal mapping surveys, and some really cool satellite verification projects. Twin Otters are used for many NOAA missions because they operate safely at low altitudes and airspeeds and are very maneuverable. This makes visual observations and sensor data collection easier. I was the first female to fly NOAA Twin Otters as the boss of the airplane (Aircraft Commander). The Twin Otter took me to almost every state in the mainland United States and up to Alaska.

After this first flight assignment, I was sent to Monterey, California, to be the Assistant Superintendent of the Monterey Bay National Marine Sanctuary. Wait. What is a sanctuary? Marine Sanctuaries are protected areas of coastal and offshore waters. The Monterey Bay National Marine Sanctuary is located along California's central coast. It spans a huge area, including pristine beaches, jewel-like tide pools, lush kelp forests, steep underwater canyons, and an offshore seamount. The sanctuary teems with life, from small invertebrates to giant blue whales. My job was to promote environmental protection, stewardship, and ocean research. This is done through coastal and open ocean resource protection, education, outreach, and research programs.

We faced problems ranging from pollution and coastal development (e.g., constructing buildings and homes on and near waterways) to wildlife or ecosystem disturbance. Knowing that people have many uses for the marine environment along its long coastline, we work with other protection groups to lessen or stop harmful effects caused by humans. While always following strict rules and permits. We also act during emergencies, enforce regulations, and educate people about conservation.

Our research, monitoring, and conservation programs test the status and health of marine species, habitats, and ecosystems. This provides critical data to support the resources we need and allows us to collaborate with a wide range of world-class research institutions. Our education and public outreach efforts promote understanding, support, and participation in protecting and conserving the sanctuary. We enhance understanding and stewardship of this national treasure through sanctuary visitor centers, public events, and volunteer and teacher education programs, among others. Building partnerships and strong public involvement is a critical element in these efforts. I moved around every few years and always took on new jobs.

"In some cases, I was the very first person to do the job. In other cases, I was the very first female to do the job."

After Monterey, I returned to flying to become NOAA's first female hurricane hunter pilot. I learned to fly the Gulfstream IV to guide hurricane forecasters on storm tracks and intensities. I was the first female in charge of NOAA's Aircraft Maintenance Branch and the Aircraft Operations Division. The aircraft maintenance branch takes care of all aircraft maintenance, both preventive and unplanned. Without maintenance, airplanes cannot be flown safely or legally. The Operations Branch manages all logistics requirements for aircraft: pilot training, mechanic training, staffing, strategy, budget planning, and implementing the plans. It also included many tedious but necessary tasks to keep the aircraft flying, the US politicians engaged, and people safe.

I retired from the NOAA Commissioned Officer Corps after twenty-five years of service as a Captain, where I was responsible for guiding the development of officers. After retiring, I missed working with younger officers to help them reach their goals and maximize their potential. I missed helping intelligent people solve our planet's science, technology, and policy problems. So, I started my own business to help companies and non-profit organizations obtain the funding and talent to perform research and provide services to the public.

What stayed the same throughout my career, and what changed from my first sea assignment to my current leadership position?

(1) My first Commanding Officer instilled in me a sense of responsibility for taking care of others, especially those with less power or who might be struggling. "We are only as strong as the weakest member of our team."

(2) I never felt confident taking on a new role but felt more comfortable each time. Pushing the boundaries of my comfort levels was critical to my growth. It was super great when I believed in myself immediately. If I did not, I faked it and sought people to help me.

(3) I always volunteered for new things, even if I did not know all the details. My ability to prepare mentally and physically, to the best of my ability, to adapt to changing circumstances, and to make decisions with incomplete information helped me immensely.

(4) I started my career wanting to have fun. I ended my career, wanting to continue to make a difference.

(5) Things were rarely easy.

(6) Persevering was always worth it.

If you are reading this, I wish you the very best of luck on your journey. Whatever you do in your life, make sure to appreciate the people around you...even the ones who do not support you. All input – positive and negative – can be used to make you stronger, better, and more resilient. I try to be a person who is a positive force in the lives of others. That also makes YOU stronger and better.

MEDICAL DOCTOR AND PHD

DRUG DEVELOPMENT IN NEUROPSYCHIATRY AND GENE THERAPIES

SANDRA LOPEZ LEON

"From an early age, I learned the benefit of following my heroes' paths and going above and beyond their accomplishments to pursue my own path and interests."

When people ask me, When did you know you wanted to be a doctor? I always respond, "On the day I was born." It was one of the best days of my life, but it was also very difficult because I was born early, and my lungs were not prepared to breathe. The doctors told my parents I would probably be okay if I survived the next 24 hours, but there was a high chance that I would not make it. Fortunately, I did! My mother told this story throughout my childhood to remind me how lucky I was to be alive. She always talked about the doctors who saved me; they became my heroes.

Once, my dad came to talk about his job for show and tell when I was in kindergarten. I did not know exactly what my dad did for work. I was fascinated when he showed up in a white coat carrying a doctor's bag, including a stethoscope (a tool to hear our hearts' sounds). My dad was a doctor! I had a hero as a father! When I came home, I told my mother. My mother told me, "Don't forget, you can do and study whatever you want; there is no profession only for men or women. You have two older brothers. You will always have the same opportunities as they do or even more. There is absolutely nothing that they can do that you cannot do."

I loved the idea that I could do more than my brothers. One was three years older, and the other six years older. I always wanted to copy them and do the latest "cool thing." At the time, girls were expected to play with dolls and boys with cars. However, in my home and at school - an International School in Mexico City - those expectations did not apply. I learned to play soccer and billiards. I opened electrical things to see how they were built and played with spiders and snakes. I did dangerous things, like jumping off a moving swing and sliding down steep banisters. I even became much better than them at playing video games. Over time, I learned that it was also cool to like things my brothers didn't, such as writing poetry, cooking, and knitting. By the time I was 15, my brothers were in medical school. They became my heroes, mentors, and teachers. During my teenage years, I was surrounded by medical students. They invited me to their classes, to the hospitals, and to their parties. I volunteered at the hospital and became friends with many doctors. Do you know you can ask doctors to follow them throughout their day at work to see if you like medicine? I watched movies and read books about medicine. I even wrote poetry about the human body and life. The more I read, the more fascinated I was about the human body. Not surprisingly, I started medical school a few years later.

"[My mother] surprised me by saying she had always wanted to be a doctor. Her dad told her it was not a good profession for women, so she studied psychology instead. That day, I decided to become a doctor when I grew up."

The study of medicine shows how each part of the body functions and how all the parts are connected, allowing your body to do all the remarkable things it does. During medical training, I got to see every organ of the body in detail. I learned about the brain, then the eye, and then the skin (and so on). I also learned how to keep each part healthy and what to do if it stops working correctly and gets sick. If you find a part of the body that fascinates you in medical school, you can specialize in this part and become an expert in it. I learned the body as a whole, which is important because everything in your body is connected. When one thing is not working, other things stop working, too.

Since everyone will get sick during their lives, knowing what is going on and what's needed to heal it is reassuring. From an early age, I learned the benefit of following my heroes' paths and going above and beyond their accomplishments to pursue my own path and interests. I have lived in Mexico, Israel, the Netherlands, Spain, and the United States. I had to learn Dutch, English, Hebrew, and Catalan. In my last year of medical school in Jerusalem, Israel, I worked in different parts of the hospital - the emergency room, surgery, and pediatrics. Since I was passionate about understanding the mind and genes, I chose to be put in a department full of doctors who were also scientists doing research. I had not realized doctors could be scientists, too! It only took me a week to decide that I wanted to be a scientist just like them. I became a medical doctor and a scientist with the highest-level degree (PhD). I work to identify the causes of diseases of the brain and mind (neuroscience) and develop new drugs to treat them.

When you study to be a scientist, you get the knowledge and tools to do the research needed to solve problems. It feels sublime to be the first person to answer a question and be the first one to understand something that no human being has ever understood before. I find it amazing that through my research, I may help millions of people who suffer from illnesses now and in the future.

"Incredibly, you can contribute to humankind by discovering the answers to scientific questions and become part of history!"

You would think that medicine and science are the most important things in my life, right? Well, the most important thing in my life is my family. I learned from my mother that having a successful career and being a successful mother and spouse is possible. She was always there for me; we spent a lot of time together. She also worked hard to have a successful professional life. I have two secrets to success. First, I always keep my family at the center of my life. I am there when they need me and enjoy every second with them. I would always pick my family if I ever had to choose between my career or my family. The second secret is to have balance in your life. If you really love any of your hobbies, continue doing them for the rest of your life.

"You must take care of your body, emotions, and social life. The more interests and hobbies you have, the more enjoyable your life will be."

Now, I would like to play a game with you. Please look at the painting below that the famous painter Eduardo Urbano Merino made for me. I am the doctor in the picture. The universe inspires the painting, but it could be the inside of a human brain. I would like you to find the stars. Did you know that our bodies are made of the same materials as stars? Yes, we are stardust.

To the right of the painting, you can find the biggest star in the sky. It is actually not a star. It is a neuron (a brain cell). The painter used his imagination, and when two "stars" touched, they became a neuron. In the painting, you can find more neurons. They are connected by chains of DNA (genes) that look like a turning ladder, also known as a double helix. My legs in the painting are made of DNA, and you can see DNA in every neuron. Do you know that all the cells in our body have DNA? It is the material that comes from our parents, grandparents, great-grandparents, and all our ancestors. It tells our cells how to make each part of our body. Thanks to DNA, our body can function; among many things, we can think, feel, and be alive. Can you believe that humans, animals, plants, fruit, bacteria, and all living things are made of DNA? That is why you can also see a plant and an apple in the painting.

The little dots at the bottom of the painting are actually numbers from Mayan culture. Those numbers are used to form a magic square. It is called a Magic Square because all sums of the numbers in each row, column, and diagonal are the same. Humans have always relied on numbers to explain their scientific results. One of my favorite things in the painting is the red apple. It can represent many, many things. You can choose whatever you like the best: the importance of eating good food to be healthy, Apple Inc., which makes phones and computers, or the apple of Adam and Eve. To end this story, I would like to talk about the white canvas on the left-hand side of the painting. It is a blank piece of paper showing that you are the only one who can write your life story. You are the one who chooses your path. So go above and beyond your greatest dreams and fly. Fly to visit the whole world. Fly in your imagination to find what others have not. Above all, enjoy your life, do things that make you happy, be creative, and live to the fullest every day.

VENTURE PARTNER

ERIN SAWYER

"I decided that I wanted to be an engineer–and use math to make the world a better place."

At the age of sixteen, I knew my calling was to be an engineer. I loved mathematics, and since many of my male relatives were engineers, I always thought engineering would be a good career for me. I'm originally from Detroit, the Motor City, and this was formative to my professional background and interests. My first experience in automotive was in my hometown, working in automotive design for a major car company. As a sixteen-year-old high schooler, rather than spending time at the mall, I traveled 1.5 hours each way to learn about automatic transmissions (automatic changing of gears of a vehicle) and experience the real work of engineering. During summers in college, I worked as an engineering intern, designing and testing diesel engines to power pickup trucks, school buses, and ambulances. I got real, hands-on experience in a vehicle's early concept and prototype phases (the first models) long before they were released to the public. I learned the amount of work it took to perfect the design of a high-performance vehicle that was safe and reliable. The responsibility was huge, but the effort of endless data gathering, 3D modeling, and calculations was worth it. These early experiences in the automotive industry laid the foundation for the rest of my professional career: finding ways to make automotive design and manufacturing more efficient.

After I graduated with a mechanical engineering degree, I continued the work I started in college and developed turbochargers (engine boosters), which were then sold to large auto companies. Turbochargers are an amazing technology that helps our cars run faster and more efficiently. I got my first taste of how a product transitions from a prototype design to manufacturing and then sold. One machine rolled off the assembly line every minute! I learned that high-volume manufacturing is stressful because everything needs to be flawless; I found this challenge thrilling.

Eventually, I moved to a company where I worked on electric cars. I felt it was closer to my core values because I always wanted to work on cars and technologies that reduce pollution and improve our environment. Since we were a new car company, we could throw out the "standard rule book" and do things in very different and unique ways. We rethought almost everything about car design and manufacturing. Combining my vast experience with this new form of energy was very exciting. I felt we made a lasting impact on the entire industry and the public's opinion of electric cars.

E. SAWYER

When I started my career, I was intimidated by being the only girl in the room. Most of my colleagues were men close to my dad's age! I felt a tiny voice inside my head say, "They don't think YOU know anything about engineering!" When I walked down the hallway, I was nervous that all the guys were looking at me, and I felt like maybe I didn't belong. While I knew that my looks should have no connection to my talents or abilities, as a female engineer, I was struggling to be taken seriously not only by my male colleagues but also by myself. My male peers would give me the "administrative tasks" of note-taking or scheduling meetings. Other times, at meetings with our suppliers, they would speak to my junior male colleagues before they realized that I was the manager leading the project. While I'm much more comfortable with my femininity today, during my early years as an engineer, I changed how I looked to "fit in" with the boys. For example, I traded my contact lenses for glasses, walked around in flat shoes, and sported a blazer instead of heels and dresses to appear more like my male colleagues.

At work, I rarely wore my long blonde hair down. Most days, I tied it neatly back in a ponytail or bun, conveniently out of sight because that little voice told me that men would never take a young woman with long blonde hair seriously. Unfortunately, when I changed my look to fit in with the men at work, I wasn't showing up to work as my true self. It is exhausting to live two separate lives, and it's draining to hide certain parts of yourself at work all the time.

As I gained more experience and achieved successful results, I became more comfortable with letting my blonde hair down, figuratively and literally. After designing diesel engine components, developing turbochargers, and managing a global supply chain of automotive suppliers, I realized that my work could speak for itself more than my looks. I began to feel less self-conscious about donning a skirt and high heels when I chose to do so.

Today, I manage thousands of car parts and do multi-million-dollar deals with global automotive suppliers. In my role, I work closely with design engineers and a manufacturing team. I'm responsible for coordinating between the two teams to bring in the components, keep inventory, and inspect all the parts to ensure the quality of the cars built. I also manage a large team of people, both women and men. In fact, I recently became the first female vice president at my current company! I achieved this career milestone by having confidence in my abilities and setting high goals for myself. I'm so proud to be a female engineer and leader in my company and prove that you can be a strong, successful woman in a traditional "man's world." When women in STEM achieve their goals, we are not only shattering the outdated idea of what a modern engineer looks like, but we're building a pipeline of young women who are vital to the future of building clean, electric cars.

CO-FOUNDER AND MANAGING MEMBER

ROKELLE SUN

"I enjoyed meeting the boys on their 'turf' and establishing friendships...those early friendships gave me the familiarity and ease among men that I needed to confidently enter a male-dominated industry..."

As the daughter of two successful professionals, my father, a professional athlete turned university professor, and my mother, an engineer, I never doubted that I would follow in their footsteps. As it turned out, though, my path was a bit more unpredictable and much more exciting than I had imagined.

When I was nine years old, my mom and I left our home in China and moved to Swaziland, South Africa. My mom, a highly accomplished mechanical engineer, decided to restart her life by opening a new business in a new country. When I was given the choice of either staying in my comfortable, familiar home or heading into the new and unknown of South Africa, I chose Africa. I had little idea what lay ahead, but even at a young age, I wanted adventure, and my natural curiosity drew me toward new experiences. Within weeks of our arrival, I began attending an English-speaking elementary school. I had never heard it spoken before. There I was, the only Asian girl in the school, with no one who could understand me and no one I could talk to. For months, I sat through classes listening to gibberish, knowing that my classmates and even some teachers thought I was dumb because I could not communicate well yet. I tried my best not to let these negative things affect me, constantly reminding myself that their beliefs about me were not based on truth but were simply a reflection of their ignorance. I knew who I was and what I was capable of, and eventually, they'd come to know that, too.

"I don't just want to know; I need to know, and if I have questions, I can't hold back. Something pushes me toward a deeper understanding, and it has served me well."

Within months, words and ideas began to make sense to me, and more importantly, I found that there were subjects I could shine in, even without a command of English. Math and science classes had a language of their own, a language I could learn well even if I didn't speak English. I wasn't naturally good at math, but I worked hard and improved my skills. Soon enough, people who judged me as unintelligent were amazed to see my ability to understand these difficult concepts, forcing them to let go of their misconceptions about me. Eventually, I learned enough English to make friends, earning a reputation at the school for my unending curiosity. Though my friends teased me about it since I was always asking questions, this natural curiosity has driven me for as long as I can remember.

Outside of school, I participated in as many activities as my schedule would allow and fell in love with ballet, cooking, art, and fashion. I also liked kart racing, horseback riding, and fencing, even though I was always the only girl among all the boys in these activities. I enjoyed meeting the boys on their "turf" and making friends with them. Those early friendships gave me the familiarity and ease among men that I needed to enter a male-dominated industry with great confidence as an adult. Besides, I began to realize that there are a thousand ways to be a girl or woman, and they're all valid. It didn't matter to me whether I was doing what people expected a girl or a boy to do or what I looked like to others. What mattered to me was finding out what made me feel happy and fulfilled and what I was capable of learning, doing, and being.

From then on, I followed the path of my own making, always looking for new ways and opportunities to expand myself. After finishing high school, I left South Africa to attend college in California. During that time, I began working as an intern at a local news station. I slowly worked my way up from a marketing intern to a production assistant, to a news writer, and to a multilingual reporter. But once I felt I'd reached the limit of my opportunities as a broadcast journalist in San Francisco, I pursued work in Paris, France, and Tokyo, Japan, reporting on the business and political news of the world. The work was incredibly fulfilling, but in the back of my mind, I knew that in time, rather than telling other people's stories, I'd make my own. I didn't know what that story would be until one day at work, when I was listening to a news story about the economic downturn, I saw my opportunity.

"...what mattered was finding out what made me happy and fulfilled and what I was capable of learning, doing, and being."

As a news reporter, I was well aware of the 2008 economic recession caused by the collapse of the US real estate market. It seemed like every newsroom story focused on businesses that had failed or were negatively affected by the global economic downturn. On that day at work, though, I heard a story describing how Japan's budget clothing brand Uniqlo® was defying the odds and thriving, even during a severe recession. A business analyst explained, "During times of economic hardship, a winner tends to take all, as consumers flock to the best value." At that moment, I thought I was ready to create my story. I packed my bags and headed to San Francisco, where I started my company, Greenland Funds, and became an entrepreneur in real estate investment and development.

Most considered entering real estate at that moment to be a terrible, or at least very risky, business decision. Many financiers were unwilling to invest in real estate during such a deep economic depression, but their fears seemed irrational to me. I understood the appeal of the San Francisco Bay Area housing market and studied economics in college. I knew that markets always changed, and I was confident that this vast market collapse was a great opportunity because the market would eventually recover.

And even though I didn't have all the skills I needed to succeed in real estate at first, I didn't let that stop me. I was fortunate to have family members with experience and a good friend / former employer who also did, just like when I was in school. As I started to take part in acquiring and renovating properties, I realized that real estate required strategy, law, psychology, and negotiation tactics, as competition for properties was high, and legal disputes against developers among city planning departments, neighborhood associations, and contractors were common. To improve my ability to face these challenges successfully, I read many books on topics in those fields. I acquired knowledge that, alongside my gentle and empathetic (able to understand the emotions of others) approach in negotiations, gave me the skills to settle deals when stakeholders disagreed on the terms. I became very valuable during negotiations and construction by resolving complex problems quickly and cheaply.

My artistic talents, such as an eye for great design, have improved my business skills. As a little girl, one of my favorite hobbies was designing and making clothes for my Barbie™ dolls. I felt the same joy when reviewing architectural and interior designs. Friends and business associates commented on my talents after seeing the homes I had built or renovated. As this part of the business felt more like pleasure than work, I sensed this was the next direction I would take in my career. This motivated me to establish my second company, Prism Capital Partners. With a decade of experience in real estate investment and development, I'm challenging myself again and bringing my passion for art and nature more clearly into my professional life. I am building design-forward, contemporary (current trends), eco-friendly homes that transform undeveloped lands and older, basic homes into creative, inspiring spaces. I'm making my mark in the world by creating innovative, functional homes (designed to maximize the use of living spaces) while providing jobs and helping create environmentally conscious communities.

My accomplishments and failures have taught me that the more I try, the more opportunities I have to succeed. Competition in real estate investment is fierce, so I've had to accept defeat over and over. I've put hours and hours into projects that never became realities, and I've lost more opportunities than I have won, but still, I'm successful. To me, that's just the way business goes. I've learned that, above all, with hard work, my skills and talents will eventually speak for themselves, and if there's something I want to do, I have to take the risk, jump in, and do it. I can always adjust the course if it doesn't work out. I believe life is a journey. We have to vow not to limit ourselves, especially not by fear of the unknown, because taking risks and making changes along the way can be an exciting pathway to happiness. Life has taught me always to have faith in myself and follow my talents and passions toward fulfilling my potential and my dreams. I hope you'll do the same!

ASSISTANT PROFESSOR

EMMA TOWLSON

"You will spend much of your lifetime on your career. So, find something - preferably many things - that is worth devoting your precious time on this earth to, and be relentless in your pursuit of it."

I could not believe my eyes. It was one of the most beautiful things I had ever seen. More than 300 young South African learners were carrying desks and chairs and arranging themselves in the large pigeon-filled halls of the community center. They could have stayed home but chose to come here and study for their upcoming final exams. Twelve of us had arrived from England a couple of weeks earlier to spend our summers working in Alexandra, a township near Johannesburg. Our aim was to teach math, reinvigorate the subject and their classrooms with our creativity, and, in turn, learn from the experienced local teachers. We were all university students at various stages in pursuing our scientific degrees, mine being in math and physics. We were part of an initiative at our university to improve mathematics education in Africa. The program stood out to me for its dedication to quality, perceptible results, and the belief that teacher exchange was a "two-way street" between the two continents. Indeed, selected African teachers were also flown to the United Kingdom (UK) for workshops. Experienced teachers and university lecturers took part alongside us. As university students, we did not have much experience, so we underwent six months of formal teacher training, theoretical and practical, before we made the long journey to South Africa. Another motivation I had for getting involved in teaching was that I was afraid. Afraid of standing up in front of a room of people and speaking.

How could I ever hope to succeed without facing and overcoming my fear of public speaking? This was my second time in Alexandra, and for the first few days, things were "normal." But quickly, rumors of a huge strike–affecting the entire public sector, including all schools–became a reality. The government was introducing a new law that would heavily affect salaries. This disrupted our work, and some days, we were sent home partway through our classes, with fewer learners coming to school.

Things came to a head within a week or so. At my school, there was a meeting with all the staff one afternoon. As was quite usual, they easily moved between the languages of Xhosa, Afrikaans, and English, so we did not understand all that was said. What sticks in my mind, though, is the moment the person leading the meeting jumped on a desk to excitedly announce to the room that, as of now, they were on strike. "You will not come to school. If you are seen near the school, you will fear for your lives, homes, and families." And just like that, schools and the entire public sector closed. Schools, police departments, fire departments, and hospitals even shut their doors, and it was not safe for anyone to be found near these places. Driving in the next morning in a last hope to continue our work, we passed lines of demonstrators in the street. They were chanting and dancing with signs around fires in trash cans. Could anyone blame them for standing up for themselves in this way when they felt threatened?

Not a soul was to be seen in a usually wildly buzzing place. We walked through the dusty entrance to the main building. It always felt a bit like a jailhouse—this concrete building formed a large rectangle with a big open courtyard in the middle. Standing in the courtyard, you were surrounded on all sides by three floors of classrooms, each floor open to the air and lined by metal bars. A lone water tap, also metal, stuck out of the ground at one end of the dusty courtyard. Drip. Drip. Drip. Ironically, a sign in the distance warned to conserve water as little fresh water was available. We started systematically working our way around the classrooms, looking for any students or anyone we could work with. Each big metal windowless door swung open to reveal empty, dull concrete rooms with broken chalkboards and furniture littered with candy wrappers and dust. We found difficulty in opening one door. We pushed harder, and something grated along the floor as the door began to open. A large trash can had been holding it closed to keep out the cold of the South African winter, and two students were huddling in the corner of the classroom. They had nowhere else to be and nothing else to do than talk.

"We arrived at the school to be greeted with the grimmest sight. The usual loud noises of rowdy students were missing; the silence felt so empty it had a physical presence."

As it became increasingly clear that the situation would not be resolved quickly, we became more upset. The grade twelve students had their Matrics – their final exams –in a few short weeks.

How could we sit aside and do nothing at such a critical moment? These exams make or break university applications and set the stage for future careers. The textbooks had many mistakes, so even self-learning was, at best, dangerous and, at worst, impossible. We also felt an acute sense of responsibility that while we weren't being paid for our work, a huge amount of time and money had been invested to train us and to bring us to this point. We had but six short weeks to maximize our impact. And so, we began to scheme. It was impossible to teach in the schools. The reasons behind the strike were, frankly, none of our business, and we had no interest in risking the program's future by mixing with it. We needed neutral ground, and we needed the counsel of the locals.

We spoke about the professional and personal connections we had made, especially with our colleagues at a university in Johannesburg. We learned of a community center in Alexandra, which quickly became a good location–it had several rooms, the largest with a capacity for around 150 people. Our very dedicated colleague had been driving us to school every day in a minivan, taking care of us, and helping with our work. He offered to continue to drive us in each morning, remain all day, and serve as a kind of "human barometer" for the local mood. At the slightest hint from him, we were to drop everything, get in his van, and go. We never felt unsafe, but he informed us that the situation was precarious and could become volatile. We called a meeting with the leaders of our project. Everyone sat around a long table, and we presented them with a detailed plan to move our operations to a different place, provided estimated costs, and asked permission to go ahead with it. We had each raised some money to bring along to fund whatever we felt would be most useful, and fortunately, we did not need much more to realize our vision. I still feel quite emotional remembering the disbelief on the leaders' faces. They were overwhelmed by our determination and how much we had already carefully thought through what needed to be done. "And you all…you all feel like this?" They pledged to provide support. That day, we headed to a big factory outlet store and used the money we had to purchase hundreds of notebooks, pens, calculators, and large pads of paper. The local university found us some stained old whiteboards and a couple of projectors. We spent all evening scrubbing the boards clean to a usable state with nail polish remover. In no time at all, we had it: a pop-up school in a van.

Because of the unpredictable nature of the situation and the likelihood of it changing without notice, we would have to arrive anew every day, set up, and clear up all traces as if we were never there. In the meantime, others phoned and texted anyone they could get a hold of to spread the word of our plans and tell them how to find us. We had gathered as many contact details of the grade twelve students as possible in those last days before the schools became no-go zones. We put up flyers and posters around the township informing them that some student teachers from the UK were opening a winter school to prepare the students for their exams. We had no idea what to expect or even if anyone would show up on that first day. We arrived and scouted the rooms to understand the space we had to work with. We wasted no time setting up projectors on sketchy electrical wiring, whiteboards, pads of paper, directional signs for each hall and topic, and chairs and tables. It soon became apparent that word had spread – students were arriving faster than we could put out chairs to accommodate them. Without question or hesitation, they joined in and followed our lead, putting furniture into place.

Around 150 attended that first day. Classes of fifty were rarely quiet or low-key at the schools, yet you could hear a pin drop in this room of 150.

At the end of the day, we all swarmed around like little worker ants – us and the learners – tidying the community center and packing everything we had brought back into the van. At the beginning of every new day, we had to do it all over again: unpack, set up, and find a way to fit all the new people who kept coming into our three rooms. At any moment, we never knew if we would have to drop everything and run. Our South African friends kept an eye on the gates. We taught everything we had the skills to teach: math, computer science, chemistry, and physics. We provided past exam questions and changed topics based on student feedback and the parts they found the hardest. Due to the demands of the situation, we rapidly developed the ability to throw all plans out the window when the unthinkable happened, make new ones on the spot, and confidently and effectively address a room of hundreds of students.

Over the next three weeks, we would teach more than 300 learners daily, helping them prepare for their Matrics. We'll never know exactly how information about us spreads, but we can hardly stop people from finding out about our school! Local reporters would turn up, and we had to be extremely sure to keep them out because we did not want to be in the news; this made us very careful around all adults. It seemed news had traveled far and fast, and we even got official written thanks from the South African Department of Education for our efforts. When it came time for us to leave, we donated all the materials we had purchased to the schools we worked in. And almost like magic, the schools opened again, and life returned to normal. That winter, the school became something much bigger than any one of us. I was so moved by everyone's unquestioning willingness, tenacity, and intelligence of the learners themselves.

They just needed someone to provide an avenue, a way, and empower them. A few months later, my experience gave me the courage to take a chance and create another such avenue – I had a crazy idea to increase our fundraising efforts. I stood in front of an auditorium full of that year's aspiring teachers and, with a fast-beating heart, asked, "Who is interested?" Every hand in that room went up.

"You see, for the most part, we're all in this together, and everyone is striving to do some good."

And that's how I found myself leading a fundraiser of sixty volunteer teachers across the UK to climb the three highest mountains in England, Scotland, and Wales in twenty-four hours. We raised a large sum of money for continued work with Africa and changed the face of sponsorship for the program. Another year in Johannesburg, I was fortunate to meet a young man in a particularly sharp suit at an event to get an international business to help the local schools. He greeted us with a big smile and told us he had been a learner in a program similar to ours. Inspired and equipped with the knowledge of how to achieve it, he dedicated himself to his studies, earned a scientific degree, and was now in a great job at this business. This one man had lifted his entire extended family out of poverty through his dedication and successful career. He was a living, breathing example of how one's destiny can change with education, empowerment, and access. It is our collective responsibility to fight for these things; the greater our privilege, the greater our responsibility. Do not doubt what can be accomplished with passion and determination; remember, we are stronger together.

SENIOR STAFF SCIENTIST/PRE-CLINICAL SURGEON

DARCY GAGNE

"...[I] work on some of the best medical devices in their class. I am surrounded by astute colleagues who are invested in helping me grow and develop; I am always excited to see what I will learn next. I am listened to and respected; I have a voice."

I vividly remember being in my sister's room on the third floor of our old white colonial house at age nine, fourth grade, when my brother asked me what I wanted to be when I grew up. I told him I wanted to be a "scientist" without knowing what that meant. I loved science and all types of animals. I specifically loved anatomy (science of the human body), but I had no idea where that would lead me. I was always curious about how and why things worked a certain way. I loved catching animals, observing their environment, and learning about what they eat and why.

I was an avid student with the goal to become a veterinarian (animal doctor) since I think "animals are the best people." My guidance counselor recommended this thing called "research." I did not know what that meant, and her explanation didn't help. I wanted to work with animals and knew there was a connection between them, but I did not have a clear vision of what kind of jobs I could hold in research. Nor did I know the questions to ask to understand better, so I figured I would follow my dream to "work with animals" and attend a local college. I was afraid to explore the larger universities where I would not know every student–or any, for that matter. About 75-85 percent of the students in my high school I had known since kindergarten, so it was an easy transition for me year to year growing up. During my senior year of high school, I stepped out of my comfort zone by attending a local university and completing four courses. I saved money by getting credit toward my college degree, and I got the experience of the college setting while still in high school.

Also, that year, I volunteered at a veterinary clinic to test out if this was, in fact, the career I wanted. I observed and participated in a typical workday and saw firsthand the types of animals and problems a veterinarian has to manage (called a "case"). I enjoyed watching the complicated cases and, even more so, those that needed surgery. I was curious about how things looked under the skin and how organs worked. I felt that surgery was some magical way to fix things. I loved learning how different animal species, like cats and dogs, were treated differently, which human doctors do not experience. It was interesting to see how owners cared for their pets;

disappointing, really. They thought they were doing a great job but often did not do what the veterinarian prescribed.

"...[if] you are shy, that does not mean you cannot stand tall and humbly but clearly show your strengths."

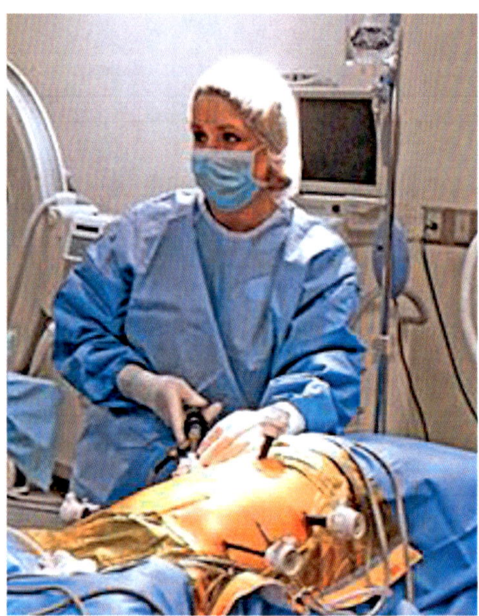

Once I started my college degree program, I learned how to handle animals properly and expanded my animal caretaking skills. I learned how to take blood for laboratory tests, give medications, and test tissue and fluids (like urine) to understand what was happening inside these pets. I was in heaven! I tried to diagnose animals and determine the treatment plan before I saw the test results. It was very technical and detailed, but I learned the science and medicine behind veterinary care. Near the end of my program, I had a chance to work (not just volunteer) in two different veterinary clinics. When I passed my exam to become a certified veterinary technician, I felt like I had entered the real world with a profession. I enjoyed it for a while but craved more responsibility and more money. The salary for a veterinary technician was, at most, minimum wage (the lowest pay permitted by law), and I felt discouraged.

Soon after, I found an ad in my local newspaper inquiring, "Do you love animals and want to make lots of money?" The clouds in my mind began to lift. I called about the role, and it was a "research" position. I figured I should explore it while thinking back to my guidance counselor in high school. The position was for a research technician at a contract laboratory (lab for hire) where drug and medical device companies pay for studies to understand more about their products. Medical devices are products that help treat sick humans and animals. I would work with a wide variety of animal species, not just dogs and cats! With my strong attention to detail and aspiration to learn, I knew I would excel. I took the job and started assisting the staff during animal surgeries. I also learned how to close surgical incisions (cuts made into tissue and skin). I was motivated by the opportunity to be promoted to a primary surgeon once my experience and skills were developed and proven on the job. I was excited and proud to use what I learned in college. I had a fantastic mentor, Vince, a veterinarian, surgeon, and PhD. I loved the job and was constantly asking questions. I sought every opportunity to assist and asked to be the primary surgeon (the lead person in charge) whenever possible. I thrived in this group as my abilities improved. I observed various medical devices placed in and around almost every part of the body and in many kinds of animals. I especially enjoyed meeting with the people who paid for the research and giving feedback to their engineers on the early versions of the devices (called prototypes). It was thrilling and made me very happy to help make better medical devices for the wellness of animals. It gave me a greater understanding and appreciation for how new medical products were developed.

At the same time, I built a house, got married, and went back to college for my bachelor's degree. My manager told me she didn't think it was possible. Maybe she thought I didn't have the strength and dedication to succeed. I don't know if she was trying to stop me, make me mad, or motivate me. Regardless of her intention, I used her hurtful words to fuel my goal of working for a sponsor company (the ones who develop medical devices), the same people I already enjoyed working with at the veterinarians'. Her doubt inspired me to learn and experience as much as possible. I remind my daughters of this story whenever they question their abilities, physically or mentally. It has taught them to persevere when they are filled with self-doubt, and it continues to give me the strength to endure difficult situations. Sometimes, I have to laugh when someone says I cannot fulfill my "crazy" dreams of doing everything at once, and I think to myself, "Watch me!"

My company is one of the best due to its strict code of ethics, dedication to innovation, quality, fantastic work ethic, research and development capabilities, and surgeon relationships. I work on some of the best medical devices in their class. I am surrounded by colleagues who are invested in helping me grow and develop. I never know what I will learn next. I am listened to and respected; I have a voice.

I am qualified to observe human surgeries where devices get implanted (placed in the body). This allows me to apply the similarities in live human anatomy to my animal surgeries. In addition, observing the preparation, instruments, and other elements of human surgeries allows me to bring those elements back to the animal operating room. It helps me provide new product feedback more intelligently, precisely, and astutely. I have traveled to Europe many times to teach surgeons how to implant medical devices. It is a terrific way to connect our animal research to the human clinical setting. There can be some discomfort at first, simply because I do not have a PhD, nor am I a physician. I "only" have a master's degree. However, once I explain that I have operated on over 500 animals and human cadavers in all areas of the body, they start to listen. Any doubt melts away once I teach them a new surgical technique and explain its importance.

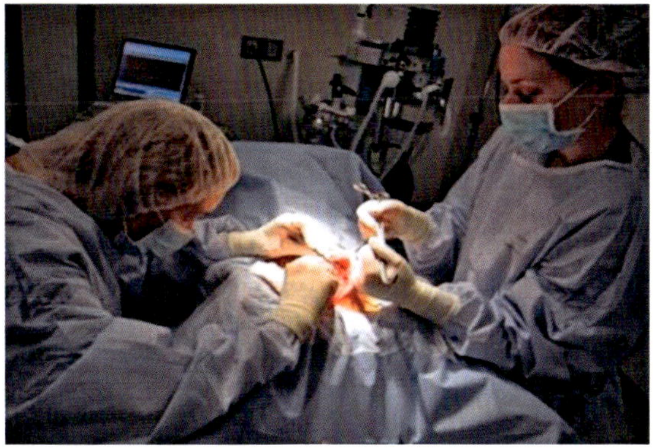

Almost twelve years ago, my manager and I were discussing how to make another important dream come true: becoming a veterinarian. He helped me realize it was not a wise move for several reasons. I am a responsible mother, wife, and homeowner, and it is unrealistic to think I could stop earning money and take on tuition debt to attend school. I did not want to miss out on my children growing up, and the return on my investment in my future salary was probably not there. After many long discussions with my manager, father, and husband, I decided to pursue a master's degree instead. Due to my potential and proven performance on the job, I received financial and personal support from my company to do so. I kept working full-time with a long commute and attended my classes. Sometimes, I traveled over 200 miles a day, then came home to parent my kids, take care of the house, be a wife, and do homework. I was not a quitter and knew I could do it. My husband cared for the children to ensure I had enough time for school and work. When he was busy, my father or my husband's parents would help. I missed out on family events occasionally but knew it would only be for a short time. My girls don't even remember this part, so I did an excellent job maintaining stability in my home life and with the right timing. After receiving such overwhelming support from my family, I was thrilled and proud to share graduation with them. Crossing the stage to reach out and grab my hard-earned diploma was such a fantastic feeling. It was also a great lesson for my daughters to see that you can achieve your dreams when you put your mind to them. Sometimes, the end goal and purpose can be easily interrupted or forgotten. I had to stay strong and focused to set an example for my children.

I am fortunate to have an incredible mentor with whom I am still friends; he would do anything to help me. I asked him once how I could return the favor. He simply said, "Pay it forward." I am committed to doing this for as long as I can. I have taught numerous animal science and research courses and interns at my company to honor that commitment. I have stayed in touch with and continue to guide some of them. Learning how often our paths have crossed over the years is an incredible feeling. Some of them have even gone on to veterinary school! I am always seeking opportunities to share my knowledge about the different career paths to research as they are not well known. I am also dedicated to continuous self-improvement and have begun a PhD program to strengthen my capabilities and my family's future.

"...do not be afraid to take risks because the most unexpected result may come from it, and if you don't try, you will never know the outcome."

It is never too late or too early to get started on your dreams. Don't be afraid if a choice you made is not working out; you can always change your course or stay committed to continuous learning and be successful. Your path will be full of twists and turns; learn everything you can at each bend! Every experience will make you stronger and more prepared for the next. Not only will you gain experience, but you will build friendships along the way. Also, do not be afraid to take risks because the most unexpected result may come from it, and if you don't try, you will never know the outcome. During difficult times, know there is a light at the end of the tunnel; they will not last forever, and you are here for a reason.

DIRECTOR

LATHA PARVATANENI

"It is easy to hold yourself accountable for all your failures. However, sometimes you need to be kinder to yourself and take time to think and reflect on your success."

My first and most important mentor was my mother. In their early twenties, my parents left the safety and comfort of their home and family in north Sri Lanka and set sail on a ship hundreds of miles to the United Kingdom (UK). They decided to leave the beautiful country of their birth because they were ethnic minorities and had few opportunities. Year by year, civil unrest increased, and they dreamed of starting a family in a land of opportunity. Their journey by boat took three long months. They had only the name and address of a family friend in London. My parents had established themselves within five years of arriving in the UK. My father was a civil and mechanical engineer and used to travel across Sri Lanka, teaching at universities. He set up his own construction business in London; my mother was working her way through school to become an accountant. They supported their family back home and brought their siblings to the UK one by one. All this while working and studying and caring for two young daughters. I always looked up to my mother as a woman of unnatural talent, creativity, and spirit. Unlike many women of her generation, she was educated and professional and worked while raising her children. She showed me the strength and drive it took to be successful.

While growing up, I remember the most important things were getting a good education and working hard. We also had a powerful sense of family. My father came from a family of 12 brothers and sisters, and my mother came from a family of five. We lived as an extended family for some time as my uncles, aunts, and cousins came to the UK and shared our house. The sense of family was vital, along with a strong sense of right versus wrong, which came from their upbringing and religious beliefs. These things set rules we all followed and how we lived our lives. They also wanted us to retain our cultural identity, so we learned South Indian classical music and dance, and they worked hard to establish a Hindu temple in south London.

As a little girl, I was always curious about the world around me. I loved biology, particularly dissection (cutting up things to study specific parts) and doing experiments. I was pretty impatient, and facts were always important to me. For many years, I wanted to become a veterinarian (animal doctor). For as long as I can remember, we would have goldfish, terrapins (turtles), cats, budgies (parakeets), and a dog at home. I don't know how my mother coped with the menagerie, but she took it in good humor.

"...have a plan, but be prepared that it may change many times. The door of opportunity may not always be open to you, but don't be afraid to push against it and be persistent."

Growing up in a family where hard work and toughness, combined with the importance of looking after others and wanting to help people, led me to decide to become a doctor. I am quite stubborn and will work hard to solve a problem, and I never take 'no' for an answer. When a tutor at my school told me that "women don't study medicine," I was even more determined to do so.

Medical training is very demanding but also very rewarding, as there are so many fascinating facts to learn. The human body is made up of a very complicated set of organs. When things go wrong, a doctor needs to be able to look at the facts and the results of tests, but most importantly, be able to talk with the patient. You can be the cleverest scientist, but if you cannot empathize (feel what others feel) with your patients, you will be of no help to them. Having finished my six years of medical training, I had to decide what to specialize in. I started training in anesthesiology, the science of using special medicine to make people sleep during surgery or when they need a medical procedure. I loved the emergencies, the drama, and the unexpected. Having a hectic job and exams to pass was not a great combination, and I failed exam after exam. I was becoming more disheartened, but my senior supervisor, who was also a great mentor, found me a quieter job working in a clinical research unit, studying new medications. I enjoyed working in clinical research, and my career took another turn at the end of my six-month assignment. I enjoyed the excitement of working on new and promising medicines - seeing the results of the studies and how the body was handling them - so much that I decided to stay in clinical research in the pharmaceutical industry.

After five years of research and while looking for my next move, I decided that although I loved the work, many of the medicines I was researching did not make it to patients (they were not approved for use), and I wanted to work for a company that could bring these drugs through research and make them available to patients. I worked on a medication to treat arthritis (a disease that makes our joints hurt) and a vaccine for girls to prevent cervical cancer. Cervical cancer is a sickness in a special part of a girl's body called the "cervix." The cervix is like a doorway between the inside of the tummy and the outside of the body. Now, I am working on a medicine to treat a very rare form of epilepsy (disease of the brain) in young children, for which there is no cure. I love my job, as there is something different to do every day. I work with very clever scientists and doctors who look after these patients. I go to scientific conferences and learn about what research is being done. Still, more importantly, I get to meet the patients who suffer from these conditions and their parents. I can see what difference these medications, the science, and the skills of their doctors make in their lives and well-being.

Several years ago, the importance of this was made clear to me when I discovered a lump in my breast at the age of 48 and in training for my first half marathon. As a doctor, I should have picked this up first; however, I thought it was merely a cyst (harmless lump). I was a healthy woman with two young sons and no breast cancer risk; I felt fine. I waited to see if the lump disappeared and carried on as usual, completing my race in record time. A few months later, it was still there, and I went to see my general doctor. Although she thought the same way I did, she referred me to the local hospital for testing. I was seen on Christmas Eve, thinking it would be a quick assessment, and I was told that everything was fine so that I could get on with my life. I was seen by the most junior doctor in the clinic, who asked me questions, examined me, and explained the next steps. She unfortunately said that I would need to have a "mastectomy" (surgery to remove my breast tissue) when she really meant a "mammogram" (x-ray of the breast). I was given a slip of paper to take to the x-ray department to schedule the mammogram.

The receptionist, who was in a rush and too busy to think about me, the patient standing before her, told me to wait and that they would be in touch to arrange the mammogram. Both of these women were qualified healthcare professionals. Still, they did not realize that while they were making arrangements for my next set of tests, I was remembering what I had been taught about breast cancer during my medical training.

With the help of family and close friends, I found a surgeon who could see me in early January. All of the tests were completed within a few weeks. When I saw my surgeon, he gave me the news that I had been both a little scared of hearing and wanting to hear: "You have breast cancer, and it is Stage 3," which meant that it had spread even more than my original tumor to my lymph nodes (where special cells that sound alarms when germs enter can be found). I was so terrified and in shock that I could not understand anything after he spoke those words. All I could think about was how awful the treatment would be for me, as well as the impact on my friends and family, especially on my two boys, who were 12 and 7 at the time. They had recently lost their father, and I was the only one there to support them. What would I say to them? What help would we need as a family? How would this affect my job, and what would I tell my employer? Being a worst-case scenario planner, I kept thinking about how much longer I would have to live. None of us deserved this, especially considering my children's misfortune of losing their father.

"It is essential to have great role models who support you and encourage you to be better than you thought you could ever be."

With my medical training, I was used to being the one making the decisions and keeping a clear head even when dealing with uncertainty. As a sole parent, I had become very good at multitasking. My mother had worked hard and raised her young children with very little help. I had help and support, but I was doing the same.

However, having become the patient, I felt I didn't have time for this. I have always been the efficient, strong, practical type who just gets on with things. Although I am relatively social, I rarely show I am scared or admit how I feel to others. Once cancer had been identified, things moved very quickly. I had been most afraid of the side effects of chemotherapy, a series of very powerful drugs used to kill cancer. Losing my hair meant it would be evident to everyone that I was having treatment. I found it extremely difficult to look strong when something was obviously wrong with me.

The physical discomfort was terrible. However, the emotional impact was worse, and keeping a positive mental mindset played a key role in my recovery. The most essential thing in such an uncontrollable situation is finding areas of control and keeping faith in the doctors. In my job, I worked with doctors, nurses, and patients.

"It is easy to hold yourself accountable for your failures. However, sometimes, you need to be kinder to yourself and take time to think and reflect."

I realized there was a big difference between what I had thought was important to a patient when I was in the role of doctor and what was important to me now, having become a patient. I got through the next six months of treatment with only a few more problems. A fluid formed and filled my arm, meaning I needed more therapy (such as massage and exercise). Much of my week was taken up by hospital visits, leaving few other things to look forward to. Marking significant times, such as the end of chemotherapy, with something special became crucial as it kept me feeling connected and positive throughout all the uncertainty.

Many years have passed since I had my treatment. It has made our family stronger and has taught us never to take things for granted. As a friend, it has made me more supportive of anyone going through illness. As a pharmaceutical doctor, it has given me a greater understanding of what a patient with such a troubling disease might feel like. I have also learned that trust in the doctors, nurses, physiotherapists, dieticians, and even wig makers is crucial. Everyone has an important part to play in supporting the patient. I took full advantage of the other treatments recommended to me, from the counseling that helped me convey my feelings to non-family members to the acupuncture treatment for my hot flushes (which became an emotional and spiritual release). Even the medical education agency (with whom I had previously worked) kindly sent me some comic books to explain the disease to my children, as well as the drama therapist who worked through activities with us and created a place where we could talk about how we were all feeling. Being a patient resets your expectations and priorities.

I recently completed my Master's in Business Administration (MBA) through a scholarship from the University of Warwick while running my business at the same time. During my coursework, I met many strong people who excelled while having a young family or newborns.

For those of you considering a career in STEM, have a plan, but be prepared that it may change many times. The door of opportunity may not always be open to you, but don't be afraid to push against it and be persistent. Everything that you do is an opportunity to learn. It is essential to have great role models who support you and encourage you to be better than you thought you could ever be. Whatever you do, always keep to your values despite what others do, and work with love and respect for others. Have lots of fun along the way because don't forget that, amongst all the noise and chaos, life is short!

SCIENCE TEACHER AND ENTOMOLOGIST (RETIRED)

ELSA SALAZAR CADE

"When you follow your passion, heart, and skills, extraordinary and sometimes unimaginable opportunities appear."

I grew up with my family in San Antonio's Westside in the 1950s with five brothers and sisters. Since my house was baby blue, with white trim and a purple front door, it was perfect for a little girl like me. We considered ourselves very American, although the music from my mom's radio was in Spanish. My first language was Spanish, but I quickly became bilingual (someone who knows two languages). Dad was a welder for the Air Force and also in the San Antonio Westside community. He was very bright and could do anything! As a master craftsman, he could weld, chisel, paint, and craft whatever was needed for his job or projects at home. He could look at something and see an unusual use for it. This helped me learn how to think "outside the box." He won much recognition from the Air Force because of this and for being a master welder. Equally inspiring was my father's mother, my Nana. She was an herbalista. An herbalista knows how to use teas and plants to heal people and make them feel better. She would apply a clean spider web to a cut to stop the bleeding. Crushed pill bugs would be rubbed into a minor skin irritation for relief, and mud was the treatment for a bee sting. The time I spent with my Nana is why Spanish was my first language, but the time I spent at my house was bilingual.

I clearly remember my mom. Tiny, pregnant, and barefoot mopping the kitchen floor. She was a fabulous cook. As a young bride, she learned how to cook my father's favorites from my Nana. Even in hard times, when we ate pinto beans and tortillas with fresh salsa, they were delicious! She would pack us gorditas (masa pastry filled with beans) for lunch. Overjoyed with our yummy meal, we would forget about our limits as kids. She kept a great home, and we all had to chip in and keep it that way. My mom loved the beauty of nature. She especially loved dogs, cats, birds, flowers, and plants. One time, to please my mom for all she did, my dad built her bird cages for her finch. My mom would say we were too many kids to take anywhere, so my dad tried to provide fun things for us to do in the back lot. On either side of our backyard was monté, wild scrub mesquite growth, with animals like rattlers, spiders, ants, grasshoppers, and birds. My dad kept it mowed for us except when the wildflowers burst with color in the spring. There were great patches of bluebonnets buzzing with bees and butterflies.

Much of my time was spent looking at them and gathering bunches for my mother. Among the flowers I picked were lovely butterflies, grasshoppers, and crickets. Life was all around me. My mother loved animals and birds, and so did I. El Cenzontle (Aztec for "the bird of 400 voices", a type of mockingbird) would sing outside my window. El Cenzontle's song is sweet but powerful. They imitate other birds and can even sound like the hiss of a man and a cat's meow. These creatures were so interesting and amazing to me that I had to go to the bookmobile (a bus designed to be used as a library) and learn more about them.

There was also a harvester ant colony in the middle of the back lot, which required me to throw a bug into the nest. If I found a giant orb spider, then, of course, she needed a bug snack, too! In those hot Texas summers, I would hear the cicadas singing loudly in the trees. I would catch a grasshopper and shove a blade of grass into its mouth (or mandible) to feed it. I would tie a string around a horned toad lizard and take it for a walk. I watched insects fly into the kids' pool my dad set up, scooped them out of the water, and put them in a jar to watch them. Then, on Monday, I would walk to the bookmobile to take out books on insects written for kids about them.

As the eldest girl, I had to help my mom with chores around the house. Afterward, I was out the door and into the backyard. Also, when the boys were too rowdy to stay inside, I was booted out with them. So, as a little girl, I often tagged along after them. Following the boys over fences meant I wound up tearing my dresses because, of course, girls didn't wear pants at the time. But that didn't stop me; over the fence, I would go. I was paying no mind to the rips I chose not to hear. It was the age of Sputnik (the world's first human-made satellite of the Earth) and the exploration of outer space.

"As a little kid, my brothers' friends called me a walking encyclopedia because I constantly recited facts about animals, plants, and nature."

I went to school in the Edgewood City school district, a district so poor it was a chapter in Jonathon Kozol's book *Savage Inequalities: Children in America's Schools* entitled, "The Dream Deferred, Again, in San Antonio." Mexican-American children faced such unfair treatment in education that it was declared unconstitutional (unlawful) in Texas. Nevertheless, I loved kindergarten and the nun who taught my class, but when I went into first grade, I had terrible nightmares over the new nun. So, my mom took me out of that school and put me in public school. I loved it! Walking to my new school, Loma Park Elementary, was an adventure, even though it was just a few blocks from my house. It was up a hill that produced red and yellow clay when it rained. On the way back from school, against my mother's wishes, I would play along the little creek that ran down that same hill. She would ask me if I went to the creek, and I would say, "Oh, no." But of course, my shoes gave me away. We also had Toritos, or ant-lions, at Loma Park. I remember staring at the pits they made in the soft dirt under the stairs next to my classroom during recess. I would look around for an ant to drop in and wait for its demise at the bottom of the pit.

A Torito would fling up dirt to make the ant lose its footing and then slide into the Torito's eagerly awaiting jaws of death. It was also delightful to scoop one up and feel the tickle on my skin as it moved backward. The natural world was my playground and provided endless fascination.

"There were many troubled children in that class. I remember my teacher putting her head down to weep sometimes."

When I arrived at Loma Park School, the principal thought I was too smart for first grade. I was put into the second-grade class with the slowest children so I could keep up with them. In effect, they placed me into what would now be called a special education class. This would come to affect me greatly after I became a teacher and allowed me to think of new ways to teach. There were many troubled children in that class. I remember my teacher putting her head down to weep sometimes. She would ask me to preach to the class about the Baby Jesus so she could take a rest. Looking back, teaching seemed like something up my alley. By the time I was in junior high, I was in the Future Teachers of America club.

When I was 14, my teacher asked our class to put together a collection of ten insects. We had to give her an old cigar box that she filled with wax so we could push our insect pins into it and mount our specimens. I submitted 50 instead because my dad took down an overhead light shade with many beautiful dead bugs inside. Jackpot! I turned it in, and my teacher was pleased. I got a perfect score on the project, plus extra credit. I loved science class, but sometimes I was bored. Not because I didn't like it, but because I was somewhat impatient with my teacher trying to get her other students to understand. Some of my classmates had trouble reading and understanding science concepts. I was bored with the repeated explanations. She tried to get them to understand. I was a bit of a science nerd and socially shy but a strong public speaker. Since I was so small and short, I had to speak up or be totally ignored. I liked the theater too, especially behind the scenes, crafting costumes and painting canvas sets using my hands. As you will see, this came in handy later in my teaching career.

That same year, the movie *Born Free* was released. My friend Pamela went to see it. I knew I never would because my parents could not afford to take us to the movies. So off to the bookmobile, I went to get the book. My first name was uncommon then, so I was fascinated by sharing a name with one of the lionesses in the story, Elsa. I loved reading about a woman in nature and the love between the woman, Joy Adamson, the naturalist, and Elsa, the lioness. How exciting it had to be camping in the African savanna! Africa was in my dreams. Around the same time, Star Trek was a new television show. While other girls were falling in love with The Beatles, I fell in love with Mr. Spock, a fictional half-human, half-alien science officer. I had a poster of him on my wall. I loved the idea of being an alien. So much so that I even thought I was one for a little while.

I was the first person in my family to attend a major university. My parents supported my older brother at community college, but they wanted me to go to work answering the phone as a receptionist for a friend of my dad's. They didn't understand that I was too shy to talk on the phone!

"Stanford University offered me fully paid tuition, but my father tore up the offer letter. He insisted I take typing, the only course I nearly failed! He said that, 'Good girls didn't go away from home.'"

I decided to go to St. Mary's University, which was very close to my house and almost within walking distance. It was also across the street from my former high school. During that first year at the university, my relationship with my former science teacher, Dr. Bill Cade, developed into more than a friendship. Bill and I married the following summer and moved to Austin, TX. He was to work on his master's degree, and me on my undergraduate degree in education at the University of Texas. Dr. Dan Otte taught my freshmen year ecology class, and I had entomology classes (the study of insects) with Dr. Osmond P. Breeland. I remember a lecture by Dr. Otte with his chalk drawings of the Intertropical Convergence Zone, known by sailors as the "doldrums" or the "calms" because of its monotonous, windless weather.

I sat in the front row every time; I was so excited to attend each class. My classes with Dr. Breeland were just as amazing as he jingled the coins in his pocket and lectured about insects in his Southern accent. He was a great lecturer in the southern dialect. He could describe insects with great humor. He spoke of things that would "just fly away like a big bird" or "gone like a turkey in the corn." He brought the insects I studied as a child to life. I introduced Bill to Dr. Otte; their relationship would become a lifelong association of collecting crickets, running experiments, building fly traps, and nighttime observations of cricket behavior, with lightning bugs flashing all around us.

While walking along trails looking for crickets, we sometimes disturbed raccoons who came up to us with curiosity and fear. Armadillos crashing through the scrub brush sounded scary until they revealed themselves to be nearly blind. Sometimes, it would be a baby armadillo, and it would rise up on its back legs to sniff the air for my presence.

One night, while blasting loud cricket songs from speakers in the backyard, a method for collecting crickets, a pale, red-eyed fly appeared at the speakers and kept coming back. This fly turned out to be a parasite that shot live maggots (immature fly larvae) at crickets singing their mating call. The maggots would dig and get inside the cricket and then emerge as pupae, killing it. We had just discovered the first example of a parasite finding its prey by its mating call.

"Science was becoming something boys were dreaming about. I remember my brothers were often asked if they would become scientists, but no one ever asked me."

What's more, this red-eyed fly, *Ormia ochracea*, had only one ear, and it was in its thorax (mid-section)! Maybe I wasn't too far off from being on the set of Star Trek after all! Since then, we've learned that this fly attacks many other insects that make sounds to communicate (acoustical insects), like cicadas and katydids. Since that fateful night that started like any other, entire laboratories have been established to study this fly. Remarkably, understanding how the fly uses its single ear to locate prey has made modern hearing aids better for people.

After graduating on the Dean's list and now with a baby in tow, I continued to help Bill collect crickets at night. I would tuck our week-old daughter into a yellow folded carrier that set my hands free to work. We walked under the street lights of the University of Texas campus, picking up the black field crickets along the curb. We would drive to the university's Brackenridge Field Station and set up our experiments. In 1992, after many years of work, we published "Male Mating Success, Calling and Searching Behavior at High and Low Density in the Field Cricket, *Gryllus Integer*."

In addition to the summer cricket research, I started a job as a fourth-grade teacher in an Austin public school. After two years of teaching, my husband finished his PhD. He sent out many job applications and finally got a job as a biology professor in Canada. It was tough to leave all my family behind, my mother and father, sister and brothers. It was far away and in a different country. We were young and figured that after a couple of years, another job would come up closer to my family. I looked forward to a new life full of adventure. We packed our scant belongings into our yellow Volkswagen Beetle, which we named Jenny, and headed north with our four-year-old in the backseat. Unfortunately, Elvis Presley died that same day, and it changed everything. We planned to stay in Memphis, but all the hotels were suddenly booked. We had to keep driving further into the night until we reached Nashville. The next morning, we had to wash really well because of all the bug bites, our welcome gift from the state of Tennessee.

Boy, Canada sure was different from Texas! I was used to spring starting in February, not June. One year later, Bill's mom and all her furniture came up to Canada! After we settled, while caring for our young daughter and my elderly mother-in-law, I volunteered at my daughter's school to give presentations on insects.

"Schools and children thrive when the local community shares its talents and time. I had both!"

I also worked as a project director for Dr. Elmer Hagley at the Canadian Agricultural Department for two summers. I was responsible for creating a collection of insects from the vineyards of Niagara, Ontario.

My team and I were responsible for the area from the community of Stoney Creek just south of Toronto to the town of Niagara-on-the-Lake, over thirty miles away. I drove my team of five Canadian students to collect, identify, pin, and label the insects. We worked in a trailer nestled in a peach orchard in the community of Jordan, right on Lake Ontario. This early work allowed me to understand the grape-growing and wine-making industry developing in the region. Before that, the native grapes, which were very sweet due to the climate, were used just for juice. Now, wine grapes were being planted. I met many grape growers as I went into the vineyards to collect and identify the insects on their land. Twenty years later, my husband, as Dean of Science, established the Cool Climate Oenology and Viticulture Institute (CCOVI) at Brock University in St. Catharines, Ontario. Canadian agriculture scientists used our insect collection to determine when to spray for pests. I was proud to have helped those same grape growers I met many years before identifying pests that would harm their vines.

After the death of my mother-in-law, I had the great fortune of landing a job teaching junior high-level science at Public School 18 in Buffalo, New York. My qualifications were perfect for the job. My degree was in education, with a specialization in science and Spanish. When I first applied, a woman ran out of the room yelling, "We have a minority!" What that really meant was that the school could then call a white person who was waiting to be hired. Buffalo had a one-for-one hiring rule at the time, meaning they could only hire a white person after hiring someone of color.

They weren't excited about hiring me, but rather the person I freed them to hire. Oh, and did I mention that I couldn't make heads or tails of the Spanish spoken in Buffalo at first? It was Puerto Rican Spanish, a dialect so different from mine that I couldn't figure out that a crying child was saying the words "school bus" in Spanish. Every day for 14 years, I drove from St. Catharines, Ontario, forty minutes one way, across the Peace Bridge to Buffalo as a border guard cheerfully greeted me.

While teaching in Buffalo, I added special education students to my regular science classes. As a student at the University of Texas, I learned doing hands-on experiments is the best way to teach science. A child can read about science or hear a science lecture, but the best way to learn science is to experience it. Well, here in the science classroom, my love of theater turned to my advantage. For example, if you want children to learn to read a thermometer, give a student some ice water and another container with warm water and ask them to record the differences. Or jam the kids tightly into a corner as an example of a solid, then let them bounce out of the class as a gas! They can roll cars down a ramp to show potential energy. Experiential, hands-on learning is a core element of my philosophy as a science teacher. Science is a way of thinking about things with special processes for doing them. In Buffalo, I thought it was unfair that special education children had to pass the same science exam as my regular students without the hands-on science experiences that are so important to understanding science concepts.

"Don't underestimate your capabilities, and take every opportunity that presents itself. If the opportunities you want don't come spontaneously, do not waste time complaining or lamenting your situation; make them happen by learning something new, meeting new people, or trying something you were always scared to do."

So, when the special education teachers in my building asked if we could try including them, I said, "Sure!" I knew that children learn in many different ways and that some needed extra help. I welcomed them into my class. It was the opposite of what happened to me as a child in my 2nd-grade class. So, with that in mind, Professor Dr. Jack Cawley from the University of Buffalo was invited to see what happens in a hands-on classroom that includes children with special needs. That led to us creating a project at the

University of Buffalo, funded by the National Science Foundation called Science for Children with Disabilities. This early work made it easier for others who followed me, including special education children in regular or "mainstream" classes, and generated a seminal research paper entitled, *Including Students with Disabilities into the General Education Science Classroom* in 2002. During the project, I was observed by researchers and other Buffalo science teachers who drew a map of how I moved about the room. Instead of sitting at my desk with kids in rows, I moved through the classroom, observing them doing their hands-on activities. I was more of the "guide on the side" instead of "the sage on the stage." I thought it was funny, given that I used to track the movement of crickets in the same way. So now I was a cricket!

At all 4 feet and 7 inches tall, I was small compared to my junior high boys. I could show a lot of anger at a naughty boy and then turn around and smile at my class after the misbehaving kid was removed. They breathed a sigh that I wasn't angry with them, too. I was crazy strict in the hallway about how my students stood in line. They had to be perfectly straight, quiet, and ready for my class. I expected as much from them as my students as they expected from me as a teacher.

But they wanted to learn. They wanted to do the activities meant for their age level. They knew to go straight to their group and be ready to start the task before them. Soon after "opening my door" to collaborate with the special education teachers, the Spanish teacher dropped in, and then the student teachers, too. With all those teachers in my class, we had plenty of adults to help manage messy projects! Having another teacher in the classroom was also nice, so I didn't have to stand on a chair to write at the top of the chalkboard. My door was also open for researchers to see my mistakes and our classroom's pet hamsters being born.

I was a lead teacher for a project in the summer of 1992, along with 80 science and special education teachers. We ran a series of hands-on activities in the science classroom that included special education students and their teachers. Based on this work, I was a semifinalist for the National Science Teaching Association (NSTA) Science Teacher of the Year in 1995. This made me one of the top ten science teachers in the US. It was a great honor to be acknowledged and recognized for my work bringing science to all children. Also, in 1993, I was nominated by New York State and the Optical Society from western New York, Michigan, Quebec, and Ontario as one of 100 teachers to be celebrated and recognized for Educators' Day.

Adding an extra teacher with special education training to help during the activities makes the hands-on instruction easier for the general child. It helps develop skills in special education children. Some special education children had issues with their social behavior. They really liked being with the other children their age, so there was a lot of social pressure to behave the best they could during that time. Some of the other children had learning disabilities that were helped by their "new friends" in the group. Building strong relationships within a team helps children with special education needs get the right science education. This education is provided by a science teacher and supported by a specialist who helps the child understand the material better.

As an example of the challenges I faced in the classroom, I remember teaching through a student's seizure as he sat at his desk. I held him in his seat while talking to the rest of the class. The kids watched him, and then he came out of the seizure and continued his work. Considering they have challenges, whatever science understandings they learn might be of real significance in their life. The boy who experienced the seizure in the class went on to get surgery and continued as a student with fewer needs. So, he fit right in with a regular class.

But once my students came in, they knew I would set them up with a bit of direction, and then they were expected to start sharing, discussing, measuring, and mixing stuff. We know that children learn best by doing instead of being talked to. I could give a brief lecture and then set them loose on their work. This way, we could do messy activities and have time to clean up before the class ended. They were excited to see what the next activity would be. About 40% of their final exam was based on this hands-on work, which really helped them. Hands-on activities are exciting but require materials and management of the student behavior during the lesson. They knew we didn't have time for bad behavior if they were going to get their hands dirty.

I was then invited to be a teacher alongside Dr. Ron Doran in the SUNY at Buffalo's master's degree program in Science Education, and I did so for several years. I was able to teach and look after young people who were studying to be science teachers. A big part of the focus was how to design the hands-on activities for the science standards they needed to meet. I was also fortunate to have amazing experiences around the world outside of the classroom. These made me an even stronger science teacher because I had such a variety of first-hand experiences across a wide range of habitats. I traveled to Sidney, the capital of Australia, then to Cairns, a city in Queensland, about 1,200 northwest of the capital, and finally to the Porongurups Mountains, all the way on the other side of the country, for a meeting of The Orthopterists' Society. The society promotes communication and research on grasshoppers, locusts, crickets, and other similar insects. While there, we visited Walpole-Nornalup National Park for the Tree Top Walk, a human-made walkway 130 feet above the ground in the Valley of the Giants region, where eucalyptus trees reach all the way up!

Tiny, superb fairywrens (*Malurus cyaneus*) danced up the bark with their iridescent blue feathers. Traveling through Australia, we saw kangaroos feeding in the open fields like deer! Some places warned of saltwater crocodiles. Looking at the water's edge and knowing a crocodile could come up at me was a rather scary feeling. We climbed to Bluff Knoll (or Pualaar Miial, meaning "great many-faced hill" or place of "many eyes"), the highest peak of the Stirling Range in the Great Southern region of Western Australia. We saw tiny carnivorous sundew plants that lure, catch, and eat insects for food. I wasn't in my parent's back lot anymore!

Through the years, I also traveled from one end of South Africa to the other, searching for insects with Bill and Dan as part of an insect survey team. Once, we flew to Johannesburg, drove to Kruger National Park, drove over 1,000 miles to the savanna of the Kalahari Desert, and then drove almost another thousand miles to Etosha Pan, a 75-mile-long salt flat. From Kruger to the

Kalahari, we encountered lions, gemsbok (a large antelope), hyenas, elephants, and baboons while searching for jumping insects. We stayed at various campsites with different environments. Seeing my first lioness at Satara Rest Camp in Kruger was a real thrill! There was my "Elsa" and her cubs in a dry sandy riverbed. What a sight! Of course, there were many other things to see, like elephants, leopards, wildebeest, ostrich, baboons, and the fantastic animals of the African Veldt (flat treeless plateaus hundreds and thousands of feet high). At Etosha Pan, we looked out towards the horizon and saw a pink line. With binoculars, I discovered it was a large flock of flamingos!

On one trip, we journeyed through Zambia, where we visited the Livingston Memorial site in Serenje, Chief Chitambo's kingdom. An obelisk marks where Dr. David Livingstone's heart was buried before the rest of his body was carried to the coast and shipped to England for burial in Westminster Abbey, London. Dr. Livingstone died there while trying to map the rivers flowing into and out of the Bangweulu wetlands. Children in small settlements cheered and greeted us with friendly waves. A group of women pounding maize also waved at us; it gave me such a warm feeling inside. I waved in return as we drove by, leaving red dust behind us.

The roads through Zambia were very rugged, with big holes we had to swerve around. On the way from the city of Chipata to Mfuzi (a town in the north), we had an electrical short in the car engine from battery acid burning a hole in the power steering cable. As I was always prepared and ready to apply my ingenious solutions, I suggested we put aluminum foil over the hole and use the wire from my spiral notebook to secure it. We would have been stuck there for days if it hadn't been for me. With the cable fixed, we could continue to the South Luangwa Flat Dog Camp. "Flat Dog" is local slang for crocodiles. I was the designated sentry at the water's edge. I would tell the mollusk scientists in our group when to hurry up and get their shells before they became Flat Dog lunch. Close to the shore, they would rake up the mussel shells from the bottom of the river to study them. This was also where the elephant known as "the Naughty One" would come and steal our oranges. I remember snickering behind our tent because it was funny to see the men trying to shoo the elephant away by clapping their hands like it was a dog. It finally left when my husband started his truck up, and the noise of the engine drove it away.

I started out like many little girls who loved catching lightening bugs in a jar, fascinated by bugs and nature. This love of nature and insects has taken me around the world. When you follow your passion, heart, and skills, extraordinary and sometimes unimaginable opportunities appear.

Don't underestimate your capabilities, and take every opportunity that presents itself. If the opportunities you want don't come, do not waste time complaining or lamenting your situation; make them happen by learning something new, meeting new people, or trying something you were always scared to do. You will be surprised how many doors open when you find people who share your passions. Remember, life teaches you lessons at every turn, so take time to reflect and learn from your experiences, good and bad. Looking for crickets has taken me from my backyard in Texas to the Kalahari, Zambia, Hawaii, and Australia. Wow, what a ride! As a young girl, I never thought I would be able to say that there is nothing more exciting than watching a king cricket female with all her spiked armor pushing her ovipositor (the organ that deposits eggs) in the sand under an ancient and majestic Baobab tree along the Limpopo River. What is the most exciting experience you can imagine?

"Don't just dream it, live it!"

CEO AND FOUNDER

CANDICE HUGHES

"...to stay motivated, I needed to continue to learn and grow. After much thought and assessing my talents and skills, I realized I would enjoy working more and likely be more successful if I started my own company."

I became interested in science in the fourth grade when my teacher brought in a fossil for us to see. I was amazed to be holding something millions of years old! Right away, I knew I had to become a scientist to study living things. There were mysteries to be solved using small clues, like the fossil I held, that needed to be explored. Gradually, I became interested in the human brain because it makes us who we are, yet we need more knowledge of how it works. By fifth and sixth grade, I was sitting in my school library reading about schizophrenia, a disease of the mind where people's view of the world and thinking are disturbed. At the time, it was not a very popular hobby for a young girl. Still, I didn't care because science, especially neurobiology (the science of how the brain works), allowed me to look into the world's most fascinating mysteries and think, "What is life?"

I guess you could say I was a science geek at a young age. Like many others before me, becoming a scientist was challenging. I had my share of challenges to overcome. My family was middle class, and my parents divorced after my first year at college. My mother, a stay-at-home mom, became a single parent. I could hardly afford to finish college, but I was really fortunate that my grandparents decided to help with my tuition. Their help, along with student loans and a job at school, allowed me to stay. After graduation, my family told me that if I wanted to continue my education in graduate school, I would have to find a way to pay for it on my own. I needed to be good at managing my money, and it wasn't easy for women then. I vowed to do everything possible to maximize my chances of earning enough money to live comfortably, and I felt science would give me this path. I researched neuroscience programs (the science of the brain and nervous system) and found that most of them were free. Some even paid living expenses if your professor could support you with a research grant (money given to scientists to perform research that does not have to be paid back). I impressed my graduate school interviewer and the selection committee with my answers to their many difficult questions like, "Why can't a caterpillar be 50 feet tall?" and my undergraduate thesis (a long research paper about your own research and the research of others).

As a result, I was proudly selected to work with the head of the neuroscience department, whose research grant would cover my costs. In return, my student's research supported his larger research goals. As part of my work, I was a teaching assistant (TA) for three years in human anatomy and neuroscience classes made for medical and dental students. As a TA, I learned how to share difficult-to-understand information clearly and effectively. It drilled the material deeply into my mind because I used it every semester to teach and answer students' questions. The practice of simplifying things became a valuable skill.

I had to work long hours and be careful with my money. I avoided buying clothes or other personal items, took public transportation, and lived with roommates to pay less rent. This experience and my earlier experiences of paying for the things I wanted made me very frugal. I worked hard and came up with creative ways to make and save money to support my family and myself.

After I graduated with a PhD, I taught college classes in neuroscience, physiology, and biology for a year. This made me realize that I didn't enjoy teaching as much as I loved solving science problems. So, I left teaching and joined a startup company in New York City to build a team to discover the next hot topics in biopharmaceuticals (medicine that is made from something living, like insulin) and healthcare. Part of my job also included planning meetings where business and science experts came together to share data and discuss new inventions and how they would change the industry. It was fun because I was constantly learning new ideas from top industry executives and scientists! I also enjoyed mentoring my team and helping them learn and grow while earning profits for my company. When I mentor a team, I encourage them to use their natural strengths. This helps build their self-confidence. My office door is always open to encourage communication, feedback, and discussions about work-life balance. Giving them direction and guidance at the right times helps them grow in their careers and perform their jobs better. I am firm but also encouraging because, as a leader, we must meet our deadlines and achieve our goals. Finding the right balance between individual development and strict goals and deadlines can be challenging.

"I have learned from a wide variety of people with many different ways of seeing the world. It has made me truly feel that I am part of global humanity."

A few years later, I saw a job advertised in an industry I was not familiar with yet: medical communications. After applying to the ad and being selected to interview, I could ask questions and learn more about the qualifications and responsibilities. It was a good match for me because it combined my love of medical science with my creativity and love of writing. Once I was hired into this role, I created training programs about new ways to treat diseases so doctors and nurses could keep their medical knowledge and skills up to date. I worked with some of the top physicians in the country, making educational slides, taking online classes, and printing mini-books, so I was always on top of the latest scientific advances.

Digging deeply into one topic after another was exciting. At the same time, I worked on building teams for large pharmaceutical and biotech companies in the US, like the ones on the Fortune 500 list. I also worked with some top-notch universities. My business acumen (know-how) grew as I learned more about how these companies ran their business and I worked with their staff. I learned how to influence others and manage a wide range of personalities. In one role, the company's CEO often came to my office to mentor and advise me. After he retired, I was selected to be part of a small team to help senior executives improve their businesses because of my vision and dedication. Eventually, the spark I had for that role burned out. I was no longer challenged; I needed to continue learning and growing to stay motivated. After much thought and assessing my talents and skills, I realized I would enjoy working more and likely be more successful if I started my own company. So, I built a company to advise biopharmaceutical companies on how to assess and understand data from their products to get them ready for review by the Food and Drug Administration (FDA) (which allows therapies to be sold). I spent months learning to prepare for, create, and launch a business, including saving money to support myself, as I would likely be without income for a while. When I was ready, I resigned from my job and launched my company. I began working more directly with patient health information and data. I had to figure out how to use the information best, which was new and fun. At the same time, I had the challenge of running a company and developing partnerships with potential customers and scientific groups while delivering the work as promised.

It is exciting and scary, just what I was looking for! I have since worked with more than 14 global biopharmaceutical companies and gotten to know and work with people in London, Amsterdam, Belgium, and India, among others. I have learned so much from people with different ways of seeing the world. Between traveling overseas for work, having international pen pals as a child, having participated in study abroad programs twice while in school, and living in an international college dorm, I genuinely feel part of global humanity. I enjoy having this big-picture view. Also, knowing that I'm helping patients get much-needed medicines keeps me motivated. It makes me feel good about my work.

Many years ago, I realized I was becoming restless for change and another challenge. I was not waking up excited about work anymore. I became interested in digital health, a brand-new field with few people working in it at the time. It combines digital technologies, like mobile phones and smartwatches, with patient health data, medical facts, and lifestyle and societal factors (e.g., income, education, occupation) to deliver personalized care. For example, a medication might be selected for you based on the traits passed down to you from your parents. This information can be used to make custom healthcare plans for you. Based on my experience, I feel that it is the future of healthcare and will enable me to help patients more directly. I created a digital health company with a team of ten people.

"Starting your own business is great if you enjoy constant change and are comfortable growing by taking risks. To have a reasonable chance of being successful, every business owner must tackle the risks they face in a thoughtful, creative way."

Together, we made an educational game to help kids with Attention Deficit Disorder (ADD) improve their attention and planning skills. I chose ADD because many of my family members have it, and I knew firsthand how tough it was for the kids and parents. I wanted to create a healthcare tool that kids would like to use. Our game had a car race that let players choose drivers, mechanics, and other characters and race against other teams. Launching a company in the tech world was the hardest challenge I have ever faced. To succeed, you have to create a product that has yet to be seen, but you also have to raise money to afford to make the product. To keep the company running, I had to make sound business decisions and grow the company by creating new products. This included presenting my business plan to a group of investors. We needed their money to keep our company going until we had a product to sell. I had to answer many difficult questions from people who doubted me.

"Staying focused and positive required a lot of bravery and determination. So, to be an entrepreneur, you need a good amount of self-confidence, an ability to take risks without extra worry, and thrive on change, often in unclear situations."

You also need to be overly positive, practical, and skeptical at the same time. Being an entrepreneur is like riding a roller coaster in the dark by yourself. It can be scary and lonely, with extreme highs and deep lows. Many people think they want to start a company but then realize they prefer a more comfortable, predictable life. To be an entrepreneur, you need to think and be a little different but not care who you are because that's what makes your business unique.

Based on my team's early success with a unique digital app that is a fun educational tool, I got funding for my company from the State of Connecticut and a Milken/Penn Venture Path Award, given to businesses to make a significant impact in education. I was awarded several other grants and many awards, including being selected as a Woman of Innovation Entrepreneur finalist. Every award and grant I won was a high point that showed me I was on the right track. It proved that I was doing something meaningful and important and kept me going.

Women creating technology-based startups find it difficult to get funding and to be taken seriously. One reason for this could be that we don't look like the image of the entrepreneur shown in the media, nor do we look like the image investors expect to see when they decide to invest in a new concept with confidence. Also, we have fewer business connections unless we have family members or close friends who are investors or entrepreneurs. Being accepted into the first trainee class of The Refinery startup accelerator program helped me gain the necessary skills, perspective, and confidence to avoid these potential roadblocks. This connected me with investors, senior executives from other companies, and other entrepreneurs. Besides The Refinery, not many groups focused on female founders' particular challenges, such as fewer connections with investors and lower chances of getting funded. The Refinery provides business training, including creating investor pitch presentations with financial assessments. Launching and leading my own startup was one of the most valuable and satisfying experiences of my career. I loved laying the foundations of a new industry and leading the creation of an innovative product that could make people's lives better. The startup is like your child, where you put everything you have into it and fight for its success.

I love being a biopharmaceutical entrepreneur because by making innovative medical therapies available, I directly impact the well-being of patients. Plus, by being the boss, I create my own career. For example, during the video game development, I would get up every day around 7 a.m. and read emails first to see if there were any urgent messages from my computer programmer, artist, or other staff members working on the game. Let's say that on this day, my programmer asked me to review a new animation sequence he added to the game. I sign in to the game's beta or "test" version and play a few rounds. Yeah, it's a hard life! While playing, I note things that are not working well, such as the animation ending too soon or a character's feet being buried in the ground instead of standing on the floor. I tell the artist what is wrong, and he has more questions, so I organize a phone call, and we work out a solution. Then, I wait for them to send me a new version of the game to review.

Later on, the same day, I attended my business accelerator program, where I am learning from finance and marketing experts. They are trying to help me become a better entrepreneur. I spent a few hours listening to a presentation on creating a social media campaign using Facebook™ and Instagram™. I talk with other entrepreneurs in my group to see if I can learn tips from them for running my company better. Then, I went home and practiced my pitch deck for a couple of hours to prepare for a meeting next week with potential investors. To add more levels to the game, we need additional funding. When I'm done practicing, I check the Facebook ads I am running to build up fans. I take a break to make and eat dinner with my husband. Afterward, I spend another hour or two checking and answering emails. I usually stop working by around 8 p.m., then exercise to keep myself healthy!

All those years studying in college and working in a laboratory helped me acquire and create exciting, well-paying jobs. I was trained to be detail-oriented and how to apply scientific, problem-solving approaches like developing an idea, testing it, documenting the results, and changing it to improve the concept. Those experiences, self-confidence, and know-how allowed me to be a visionary throughout my career. Entrepreneurs who use science to create products that change our lives are in the news every day. People now recognize how critical they are for making our country successful. Globally, scientists are finally being recognized for this vital work.

I am inspired by the fantastic things other entrepreneurs are creating and the potential for new inventions. I wish I could create or implement every innovation I think of. Still, I can only work on a few ideas at a time because of the enormous amount of time and money it takes to bring an idea to life as an actual product or service. It takes a lot of effort along a difficult path to get a product on the market and into customers' hands. There are many opportunities to make mistakes along the way, from the very beginning when the idea first pops into our heads, all the way through creating a prototype, testing it out, and then making changes to it over and over again until it is final.

"I had to respond to many difficult questions from people who doubted me. This took a great amount of courage and determination to stay focused and positive."

Next, I would love to partner with a group of people to grow a startup company to a mid or large-size company, and I feel like I created something permanent to improve people's health. I also want to help women succeed in their careers and their ventures. Now, I'm helping other startups with their business and funding planning. I especially enjoy helping women-led startups. In the future, I look forward to being on the board of directors of businesses to boost their growth further and keep them on track to be successful. Often, the path ahead seems full of hurdles. Be strong but flexible so you can change your direction when needed. Don't give up; the view is worth the climb.

ASSISTANT PROFESSOR

ADRIANA L. ROMERO-OLIVARES

"As cliché as it may sound, I didn't choose science. Science chose me. If you belong to a small group of privileged people who get to choose science, go for it! For the rest of you, this is my STEM story."

A. L. ROMERO-OLIVARES

I grew up in a working-class Mexican family. Scientists were only in movies, and they were always men. When I was young, I never even considered careers in science. I did not know they were actually a thing. But I did know that I cared about the environment. I always cared about pollution, keeping our oceans clean, and protecting wildlife. Despite all odds, I have a PhD in ecology and evolutionary biology. I am dedicating my life to understanding how ecosystems work and how we can protect them.

My interest in science was planted unintentionally by my dad. My mom was the main provider for my family, so she was always working. My dad had Sundays off, and my mom didn't, so we spent many weekends together—just the two of us and then with my two brothers later on. Very early in the morning on Sundays, my dad would wake me up, and we would go somewhere outside the city or at least to the park. I hated this. I did not want to spend my Sundays outside in the heat, but I had no choice. These days were long. I was hiking, exploring, being scolded for not being adventurous enough, and eating boiled eggs and potatoes.

After a long day outside in the heat, I was dirty, sweaty, and cranky, but I felt powerful and unstoppable. Usually, we would go visit my grandparents right after, and my "abuela" would scold me because I was not "lady-like,"; "Girls don't play in the dirt," and "Girls don't sit like that." The worst one, "Calladita te vez más bonita," which translates roughly to "You look prettier when you're quiet." Mind you, my grandmother was not "lady-like" and surely not quiet; she was stubborn and tough and quite scary, to be honest. How conflicting! My dad pushed me not to be "lady-like," and my "abuela" scolded me for the opposite reason, even though she was not like that, to begin with! I would just roll my eyes. Toward the end of the day, we would pick up my mom from work, and I felt safe again. This was my favorite part of the day. My mom was not critical or pushy. She would allow me to be myself, whatever that was at the time.

From a very young age, I did not "fit the mold" of how a woman was supposed to behave. I was not delicate or feminine; I was headstrong, opinionated, and smart. I was not the best at school; I did not get all good grades, but I was good at science. I was especially good at biology, and above all, I loved the Earth. I remember being moved to tears when I heard the song "The Colors of the Wind" in the Disney movie Pocahontas, in which a Native American woman shows a colonizer (someone who came from another place and took over the land) white man who every rock, tree, and creature that inhabits the Earth "has a life, has a spirit, has a name." To me, this song represented what I felt: respect for the Earth and an urge to protect our environment. As a girl in the city, I wasn't sure how this was done. So, naturally, I started to scold family and friends if they littered, wasted water, or threw rocks at stray dogs. But I still needed to face a very important challenge: What do you study in college when you want to protect the environment?

My mom was a telephone operator, and my dad worked in construction. My grandparents were farmers and teachers. The career people I knew worked in banks or were physicians. I honestly did not know anyone who went to college to study how to protect the environment. When I talked to my parents about this, my dad told me I was brainwashed and should go to beauty school because hairstylists always had a job. After all, hair never stops growing; people always need to get haircuts, and I would always have income. My mom was shocked and embarrassed because she could not understand why I was not interested in a more "normal" degree, like nursing, social work, or business. However, she supported me because, after all, the one thing she always stressed was that I should go to college and get a degree. I reached out to others in my family to discuss my interest in a college degree to protect our environment, and I got the same reaction from them as my parents. I was confused and disappointed and thought I might get a more "normal" degree. I started exploring a few options but was not attracted to any of them. I truly wanted to dedicate my life to protecting the environment. I reached out to a family friend I thought was in a related field. He was an agricultural engineer. His advice was, "Why don't you study biology?" My mind was blown. I did not even know that there were careers in biology. After all, that career was not an option at the universities in my city, and I never thought of moving out of town for college. I immediately started researching how to make it happen. The closest university with a biology major was 700 miles from my home. I had good grades and could apply for a few scholarships to afford college. I had a plan. I presented it to my parents. They were not as supportive as I wanted, but I was determined, and they knew it. I told them, "If I don't study biology, I'm not going to college at all." This was definitely a teen drama ultimatum that I'm not proud of, but it got me to where I am now.

I moved away to college and faced new challenges, like living alone, being poor, and being homesick. I missed my family and friends but was on track to get a degree in biology. However, it was still not clear how I would use my degree to reach my goal of protecting the environment. My professors were my role models, so I thought: Should I be an activist? A science teacher? I want to say that I clearly understood the path I would take, but that would be a lie. Was a biology degree a mistake?

Halfway through college, I became interested in molecular and microbiology and how we use molecular tools to study microbes and protect the environment. I also enjoyed university life and looked up to some of my professors. Some were activists, others policymakers or passionate teachers. At the same time, they were all invested in learning and protecting the environment. That was it. It was clear. I wanted to be a professor, too.

I asked them for advice on how to be a professor like them. The answers were scary. Me? Getting a PhD in science? Five to ten more years in school after college?! Doing research? Becoming a scientist? Like those men in the movies? It sounded so scary. Slowly, with the support of my professors, family, and friends, I got into a master's program and then a PhD program. I became the first scientist in my family. Wow!

A. L. ROMERO-OLIVARES

"With the support of my family, friends, mentors, and teachers, combined with my enthusiasm, passion, and will to achieve, I fulfilled my dream of dedicating my life to protecting the environment."

Now, I, too, am a scientist, activist, and teacher. I am involved in protecting our environment for future generations and making science available to everyone. I do not want a world where young girls think that science is only for men. Reach out to others for advice and encouragement, and remember that there is no mold for what a scientist should look like or what type of background you should have. I am a scientist; I am a woman; I am Mexican. I am not lady-like. I have strong opinions and can be pretty stubborn and headstrong at times. I remember that the traits people used to criticize me for are actually the traits that made me who I am today and pushed me to be successful. My story is not about becoming interested in science at a very young age after doing an experiment in my school or after spending hours with my parents in their lab because those stories come from privileged students. Inspirational science stories of poor people or people of color are like mine. Be unstoppable, even if you do not fit the mold.

BOARD MEMBER

CHARLOTTE SIBLEY

"One of my early bosses was a legend in the industry for being very tough. She told me that I wanted too much to be liked, and she was right! I was troubled by this because it seemed like it was going to be a career-limiting factor, so I kept trying to figure out an alternative."

I have always loved math and science. In sixth grade, a small group of us were bored, so the teacher started us on algebra! This was a few years advanced for us, but we were lucky to get a jump start on middle school. In eighth grade, my science teacher was my sister's best friend. She was always "Mary" to me. Then she was my teacher and became "Miss Leavitt." She loved science and was a great teacher. She taught us not to be afraid of science and math. I didn't realize until I was much older that it must have been difficult for her to get a master's degree in biology in the 1950s. We talked about this later as adults, and she only said, "It wasn't easy." People tried to discourage her as it wasn't a "proper" subject for a young lady to study! I might have pursued science except for two things: 1) girls then were pushed toward nursing, teaching, or being a flight attendant; 2) I have monocular vision (one eye can see much better), which makes looking through a microscope with two eyepieces difficult.

I took Advanced Placement courses for chemistry, physics, and calculus in high school, but my real talent was for languages. Latin and, later, French came easily to me. I learned the vocabulary quickly and have a good ear for proper pronunciation. So, I majored in languages in college. I planned to get a PhD in French literature and become a university professor. In the fall of my junior year of college, the "personnel department" (as human resources were called then) from the United States Trust Company of New York on Wall Street came to my college campus to hire summer interns for the new "international team." The fact that I was studying foreign languages was international enough for them to consider me for a summer job - even though I hardly knew the difference between a stock and a bond! I was invited to New York City (NYC) for an on-site interview. I was engaged to be married to a student at Columbia University at the time and planned to live with his parents in New Jersey. So, I was looking for a job in NYC that summer. The weekend of my interview (and job offer!) was the weekend we broke up. But I decided to go to the Big Apple anyway!

I was the first woman summer intern and the only woman summer intern for the three summers I was there. I didn't see this as unusual at the time. I did notice my assigned table at lunch (served in a dining room!) was with the "girls," that is the secretaries, and not with the five other interns, all of whom were male. I mentioned this to my boss, and he got me reassigned, not to the table with the male interns but with the bond and stock clerks, mostly young men. I learned a lot from them. The other interns were all heading to graduate school for a master of business administration (MBA). I started dating one of them, and he kept talking to me about the benefits of an MBA: the strict approach to making decisions, the financial training, and especially, the job opportunities at graduation. I heard of Harvard Business School, but I thought it was a master's degree in accounting... and I knew that wasn't for me! By the end of the summer, he had convinced me to apply to business school. The University of Chicago offered me a full scholarship for the first year because they wanted more women enrolled to add diversity to the student body. Over 95 percent of the students were male, most had engineering or accounting degrees, and many were from the military. This was also when business schools were just starting to understand and research leadership as its own skill and the "soft skills" necessary for good leadership, many of which were in my literature and social sciences studies.

Towards the end of school, I was interviewing for full-time jobs with companies that sent representatives to the campus. I was attracted to the announcement for Pfizer, a company known then more for chemicals and cosmetics than for drugs. I was impressed with their interviewer. He also majored in liberal arts and felt it was good preparation for the corporate world. By the time of my follow-up interview at the New York headquarters, I was excited by the pharmaceutical industry! My final interview was with a vice president who also got his MBA at Chicago. I thought this was a good omen and the right company for me.

Pfizer was not well-known then, and the drug industry was not a target for MBA graduates. It was not seen as "prestigious" as investment banking and consulting jobs. Many people thought you had to be trained in science. One of my classmates said, "You're joining Fizzer...what kind of a company is that?" And I said, it's PFIZER, and it's a pharmaceutical company. He looked puzzled and said, "Farma? Like agriculture?" And I said, "No, pharmaceutical, as in drugs." His response was, "Why do you want to go to a company like that?" I said, "Because I'm really interested in the steak vs. the sizzle (substance versus flair), and the science intrigues me."

I joined Pfizer as a market research analyst, where I gathered information about unmet needs from doctors and nurses. I started with a drug to treat schizophrenia. I loved listening to the doctors describe their medical practices and patients. We didn't interview patients back then because patients had almost no say in their treatment and care. I also loved understanding the reasons for doctors' choices. What did drugs need to do versus other treatments for doctors to use them? I learned many key lessons: 1) when trying to understand data, come up with an idea of what you think the data will show (have a hypothesis), but keep an open mind; 2) analyze all the data, not just the data which supports your point of view; 3) be able to communicate your findings well (tell a story) so that everyone can understand them and 4) develop recommendations for the future research or decisions.

After a few years at Pfizer, I returned to Wall Street as an analyst for the drug industry. I recommended buying, selling, or holding stocks for large "buy side" investors like banks and insurance companies. I was the only female drug analyst on the "sell side," although many women drug analysts were on the "buy side." This allowed me to see the industry as a whole and understand what led to a company's long-term success and survival. We use the same factors today to assess the strength and future of a company. My next job was in market research for consumer-packaged goods (like Lipton® Cup-a-Soup). This was "classic market research," I asked consumers (mostly women) about their use of food, personal care, and home products. I did not understand them because I couldn't relate to the products: hot instant noodles rehydrated into glutinous, salty clumps? Now, I knew I was not the typical consumer. I was single, lived in an apartment in NYC, and I went to the opera. And the noodles didn't even taste good to me! So, I returned to the science and healthcare world to work in market research at Johnson & Johnson. As it turns out, product testing with nurses and doctors in hospitals was not that different from working with homemakers. So, my skills were more transferable than I thought! However, when my boss resigned and I realized I would not be promoted to his job, I began looking for another one. At the same time, changes were coming to how hospitals were paid, and roles and responsibilities at my work were changing. Within a year, my department was broken apart, and people and roles were changed and placed in other parts of the company.

I decided to join Medical Economics, publisher of the Physicians' Desk Reference (or PDR, a summary of information on drugs). At that time, the PDR was available only to physicians. Since I had access to it through my job, my friends always asked, "Ms. Doctor, what are the side effects of the medicine I'm taking?" This was my first real management position. I led a team of fifteen people. I had a supportive boss and a great team reporting to me. However, publishing does not have money to spend on thorough research, which was close to my heart.

A member of the newly formed Healthcare Businesswomen's Association gave me a call. She told me about a job opportunity where I would be in charge of global market research with a different drug company called the Squibb Corporation. I was excited about the job because it was worldwide, and I had not had that level of responsibility before. With a much larger budget and the ability to hire new staff, the role was attractive. I happily accepted. Shortly afterward, a PhD chemist on my staff discovered that a competitor was way ahead of us in developing a similar drug–and the data showed it was better than ours. Since this came about through extensive research, I used it as a reason for starting a competitive intelligence team. This would be the first in the industry!

My job made it through a merger with Bristol-Myers, but after that, I had to deal with seven different bosses in just four years. I had two of these bosses twice, with no promotions for me or my staff, and they even stopped us from hiring. I was (finally) promoted to a new role heading up market research and analytics for the US pharmaceutical team. This was the largest and most profitable in the company. By this time, "managed care" was a major force of change in the industry. With managed care, patients can only see a select few doctors. And those doctors could only prescribe certain medications. Also, with managed care, data on products' safety and efficacy (how well something works) was still necessary – but not good enough. So, we teamed up with our scientists to convince the marketing department that benefits (like improved health), not just features (like being easy to swallow) of a drug, were important. We also had to demonstrate "value" in patient benefits (having a pill that is easier to take than having to go to the doctor for a shot) and better health (like lower blood pressure), so our jobs as market researchers became even more important.

Since the pharmaceutical industry was changing from doctor-driven to managed care-driven and with direct-to-consumer (DTC) advertising, we were becoming more like the consumer products industry. I always wanted to make market research a key part of decision-making for drug products, just like the consumer products industry. These trends have made our jobs more difficult. Now, we had to consider the attitudes and behaviors of the managed care organizations, consumers, and physicians. Many times, they were different. It was still about the steak (substance) and not the sizzle (flair). We had to understand how we were better than the competition, focus on our benefits, and be able to communicate them crisply and clearly!

One of my early bosses was a legend for being tough. She said I wanted too much to be liked, and she was right! I was troubled by this because it seemed like it would slow down my career. So, I kept trying to figure out a different way. To be hated? Do any of us really want to be hated? To be feared? People will perform under fear but usually try to get away as fast as possible. To be respected? Yes, absolutely! So, I decided I was happy with being liked and respected. If that didn't work, I would always be able to 1) Take pride in my work, 2) Take pride in my team, 3) Know that I make a difference, and 4) Have some fun! That has kept me grounded as an employee, a manager, and a leader. It reminds me of how I got here and leads me to an acronym I'd like to share with you. PRIDE: Persistence, Resilience, Integrity, Do Good Work, and Energy. Persistence is not giving up. It's the ability to see another way, another path. Persistence is having the self-confidence and strength to keep going when the going is rough. Resilience may be one of the most important qualities in a leader. Resilience is the ability to bounce back, roll with the punches, and not get crushed. We hear about changes at work every day. Change is fine if you're the changer, but not so much fun if you're the change! Through all of it, try to keep a sense of balance in your life and a sense of humor.

My resilience was tested when I was fired from Bristol-Myers Squibb after eleven years of service. They said it was a performance issue. In reality, it was a "fit" issue. I had spoken the truth and challenged the president about an important prediction. I did it in private, carefully, but still in opposition to my boss. In the end, my group's forecast was right, and the finance and marketing departments' forecasts were wrong. My boss read me the list of how I had failed, mostly untrue, but they wanted to "make an example of me." Deep in my heart, I knew they had done me a favor because I did not want to be a part of that culture. But for someone who wants to be liked, it really hurt. Two years later, I was offered "the job of my life" at a biopharmaceutical (biotechnology and drug) company, Pharmacia. I led the global analytics, research, insights, and forecasting groups. It was the first company to have all of the global data, research, and analysis teams in one department. This gave us more information about the competition, our customers, and the market. This enabled us to make better, fact-based decisions. We did so well that Pfizer bought us!

I love the idea of making a difference by lengthening and improving patients' lives and creating engaged, productive teams. I am grateful to have worked in an industry where the mission is so important. I now serve on the boards of directors of many life science companies while still mentoring many people. I love board service–it's another way to continue making a difference in patients' lives and in the industry.

"Aim for progress, not perfection. Things don't always happen the way you want - and it's OK!"

CONSULTANT IN AVIATION TECHNOLOGY AND OPERATIONS

KARIN HOLLERBACH

"Choosing a career in STEM doesn't have to mean leaving part of yourself or your passions behind."

I've always been interested in STEM because of my curiosity about the world around me. When I was a little girl, I loved math and science and wanted to become a biologist. Specifically, I wanted to become a naturalist or field biologist. I was inspired by stories of scientists working with animals in Africa or studying the ecosystems in California, Alaska, and the Arctic (the latter being my home and where I spent a great deal of time outdoors). Since my greatest interests were biology and spending time outdoors, what could be better?

By the time I went to college, I was still spending a great deal of time outdoors but had revised my career goals to something more practical. To be honest, with less chance of me ending up unemployed. I shifted my focus to molecular biology, which could be done in the laboratory. I loved it; at the same time, I learned by experience that I did not want to spend the rest of my career inside a laboratory. While attending MIT, I also discovered engineering. This was one of the turning points in my life and career and was the result of my environment. Although I was at a university that excels in the life sciences, MIT is really more broadly known as an engineering school. I thought, "While I'm here, I should really learn what an engineer is." and promptly fell in love with it.

I ended up studying both subjects. Life science was all about understanding what is out there in the world. How can I understand what I see in the world around me? How can I make a mental model of it to understand it better? Engineering was all about designing and building. If I have a mental model of something, how can I turn it into something real? I loved those seemingly opposite ways of thinking about life. More importantly, I realized that forming connections between apparently unrelated areas really makes me light up. Although, at the time, my thinking was still mostly confined to STEM.

It's always bothered me that many people put people into either/ or categories: Either you're an engineer or a scientist. You're either a STEM person (or a techie) or you're...something else. Often, these labels are used as if you can only be one thing but not another. This type of thinking is harmful and, in a very personal way, was limiting me and what I wanted for my life.

One of my challenges has been combining my STEM interests with the other-than-STEM parts of my life. I don't just mean that I "do engineering" and like going to the theater. I'm talking about integrating my passions and developing my career more holistically. Beyond science and engineering, my passions include flying, endurance sports, adventure (especially in faraway areas or extreme environments), and aikido, a martial art, but for me, a meditation practice. How do these relate to STEM? In many ways, they don't–but they relate to me, the whole me!

I have started adding many more of my passions to my ever-changing career path. Will I add all of them? Maybe not, and that's not the goal anyway.

However, I am changing my work focus to add more of my main interests. How can I add flying to science and engineering? Although I don't want to fly airplanes for a living, I can use my understanding of flying to work with technologies applied in or by aviation. Like using planes or drones to put sensors in place and using that data for projects we are working on.

Another example is: How can I combine my love of endurance sports and STEM? I'm not interested in designing traditional products for athletes, like running shoes or bicycles. I am interested in creating wearables to improve athletic training and human performance and testing them on myself and others. Despite the stereotypes around athletes and engineers, it's possible to be both an accomplished engineer and a high-performing athlete. What about extreme environments? Traveling to these places used to be simply a hobby for me. Now, I've broadened my interests to technologies for extreme environments and will field test them, perhaps even gathering data to help solve important scientific questions while I'm there. At the moment, I'm looking at gathering data to measure sleep at high altitudes (on mountains).

I'm also interested in laboratories that develop wearables to monitor body functions, like blood pressure, during an "ultra" sporting event or expedition. One of my goals is a polar expedition, where I use wearables for bodily function (physiological) and environmental monitoring. I'm interested in the Polar Regions because they are so severely affected by climate change that they are becoming a precious place to study before their ice masses disappear. All of these interests present research options for me. Bringing together people with different backgrounds and interests opens up a world of opportunities for new discoveries and innovations. This is incredibly valuable in today's world, where there is a growing need for products that can do many different things and are at the forefront of technology. Adding these diverse perspectives creates opportunities for creative solutions.

I have built skills in my hobbies that I can bring into my STEM career. For example, my meditation practice improved my ability to focus deeply. It expanded my awareness and curiosity across all of my senses, which are excellent qualities in scientific and creative work. It has also made me more empathetic, an important trait for a leader and member of any team in STEM or other fields. Similarly, my love of flying, which also builds on focus and awareness, required me to learn the science underlying weather phenomena and the engineering related to aerodynamics and aircraft control. Even though my career has nothing to do with the weather (meteorology), learning about one STEM discipline enhances my understanding of another. In addition, flying has led me to work with others who share that interest and become part of teams working on projects much more significant than anything I can do alone. Of course, not everything I love has to be turned into something that earns me an income! But, my skills in areas that might not seem connected have made me a better leader and contributor to my STEM journey.

Even more significant, as I reflect on my life as a little girl, I realize I'm only now returning to what excited me the most: being outdoors, science, and math. Back then, I thought that these interests meant becoming a field biologist. Somewhere along the way, I allowed those interests to get separated from each other. They were part of my life, but when I was into one activity, I felt like I was leaving a different part of me behind and not bringing all of me to the moment. I realized that since they are all significant parts of who I am, I must integrate them for my well-being. Combined, they give me a unique set of experiences and skills to contribute to my work.

The little girl in me had a lot of wisdom. It's just taken me time to acknowledge that and realize that there are many STEM paths to choose from while expressing the full diversity of my life's passions.

"Don't accept limitations imposed on you by others, or even by yourself. Your career is YOURS to do with what you want."

DIRECTOR OF REGULATORY POLICY, INTERNATIONAL AND HARMONIZATION

FATEMEH RAZJOUYAN

"Chase after experiences that can help you appreciate the value of your goal and, ultimately, your purpose in life."

My story begins in Iran, my birthplace, where, at a young age, I saw Iraqi soldiers invade my country. Because of that, I faced a lot of pain and suffering. Seeing the struggle between life and death up close ignited an intense passion in me for public health, to restore some sense of balance to the world. I never imagined I would actually have the chance to bring this dream to fruition.

As a child, I was a keen observer, always eager to learn and ready with a series of questions to satisfy my thirst for knowledge. But, growing up in a male-dominated society with a government that ignored our most basic human rights, I did not have the stability and confidence to build a proper foundation for the future. However, I had a solid and secure family environment. I was fortunate to have parents who put education above everything else, a rare privilege for a young girl in Iran.

Despite this, I knew I could not have my own voice in that society or even be known and accepted for my values, thoughts, and beliefs just because of my gender. This reality hit me hard one day in high school. When I was just a freshman, I vividly recall an incident with my math teacher, who happened to be an older man. His hurtful comments really got to me and briefly made me doubt my dreams of pursuing higher education. It was a challenging moment. Intrigued by logarithmic functions, a mathematical way to handle and describe very large numbers, I found myself confused by the concept. I asked my math teacher, "Could you explain if power and logarithm are inversely related to each other?" He heard my question, paused, and stared at me. His face twisted up in annoyance and frustration. He shouted, "Why do you always need to know everything? Don't you understand that as a woman, you'll end up washing poopy diapers... I can't stand wasting my voice and energy on the likes of you..." This incident was a chilling reminder of the constraints I faced in society, where my ambitions and aspirations for a public health career were seen as abnormal simply because I was a girl. My dreams of advancing in education and improving public health seemed utterly unreachable. That math teacher made me feel hopeless, but my mother lifted me up with her encouragement, optimism, and positive energy. She said she'd stick with me and do anything in her power to give me the future she couldn't have. "Never fear dreaming big, but dream with goals, and promise me that you'll never fail to plan to realize your dreams," she said. She filled me with the courage to persevere.

At 17, my mother, father, two siblings, and I immigrated to the US, where I faced the challenge of learning English and a new culture. I started an entirely new life at an age when my mind was open to things that were different. I was driven, and mastering English within my first year of arriving became one of my biggest priorities. I set quarterly goals to increase my vocabulary and improve my reading, writing, and speaking skills. I carried a bilingual dictionary (two languages) everywhere I went. I wrote down new words and phrases in a notebook as I came across them and made them into flashcards to study later. I practiced my pronunciation by watching sitcoms and imitating the sounds and words using a mirror to shape and move my lips into the correct position. I even used 1-800 numbers printed on the back of household products, such as toothpaste, as a way to talk to someone in English. I attended English as a Second Language (ESL) classes at Northern Virginia Community College (NVCC). Before transferring to a four-year college, I planned to complete my first two years of undergraduate school at NVCC.

"I was fortunate to have parents who put education above everything else, a rare privilege for a young girl in Iran. Despite this, I knew I could not have my own voice in that society or even be known and accepted for my values, thoughts, and beliefs just because of my gender."

Life in the US allowed me to pursue my education freely, without the stress and hurdles of war and oppression. I felt so free, like I'd broken out of some invisible chains. I was liberated, emancipated. In my newly adopted homeland, I chose my friends and others around me wisely. I knew if I picked them carefully enough, they could empower me. I specifically surrounded myself with people with a strong desire for self-improvement, those who could inspire me to be better and persevere.

I am happy to say I excelled at NVCC, and with my tenacity and hard work, I became proficient in English. As my mom always says, "Hard work works!" At the same time, I began volunteering at INOVA Fairfax Hospital, which allowed me to serve my community and improve my interpersonal and communication skills. Little did I know, but that moment marked the beginning of my journey toward a career in public health.

Although I was unsure of my college major, my curiosity about everyday electronics and household items pointed me in the right direction. The old pendulum clock at our home, in particular, fascinated me. As a child, I could sit and watch how the pendulum swung side to side for a long time, trying to imagine the gears inside that led to the passage of time. I can close my eyes and still hear the steady tick-tock echo in my mind.

With encouragement from a science teacher at NVCC, I pursued Biomedical Engineering (BME) at George Washington University (GW). While interested, I hesitated to apply to GW because tuition was expensive. My mom always says, "Shoot for what you intend to hit; you won't know you'll succeed unless you try." Motivated by her words and constant support, I applied. I was thrilled when I was not only accepted but also offered a scholarship, meaning I could now afford to attend GW.

Even though I had a tough engineering course schedule, I didn't lose sight of my love for public health. I enjoyed mixing engineering and medicine through my BME research in the detection of breast cancer using infrared thermal imaging (sensing heat differences and turning them into colorful images). This showed me how BME can be used in real life. It was a challenging but enlightening journey, combining basic science and engineering to solve public health issues. It was quite rewarding and made me even more committed to my chosen path. I worked very hard and graduated Magna Cum Laude, a special distinction awarded to high-achieving students. This proved how determined I am and how much I love learning. It shows that I'm still the same curious and eager learner I've always been!

My college degree and passion for public health led me to an internship at the Office of Science and Engineering Laboratory (OSEL) at the Food and Drug Administration (FDA). This let me continue my research on breast cancer while working in an organization devoted to public health. My internship focused on Electrical Impedance Tomography, another way to detect breast cancer early. This technique makes images of breast tissue without using needles or drugs. It is being studied as an alternative to screening mammography (the most common way we currently image the breast). After seeing my mother undergo her annual mammogram, this research resonated deeply with me as I wanted to make the experience better. This technology overcomes many shortfalls of current mammography, like radiation exposure, which itself can cause cancer.

"I chose my friends and others around me wisely. I knew if I picked them carefully enough, they could empower me. I specifically surrounded myself with individuals with a strong desire for self-improvement, those who could inspire me to be better and persevere."

At this point in my life, my volunteer and research experiences and my BME training helped me define my professional goals. I found great satisfaction in problem-solving, and my engineering skills allowed me to handle and let me solve even more complex problems. This was a powerful feeling of accomplishment. What was missing, however, was a path to making an even greater impact on public health. This feeling of emptiness showed me where my true happiness in my future career lies - in assisting patients in accessing safe, groundbreaking medical devices. This led to my work as a lead scientific reviewer at the FDA's Office of Device Evaluation (ODE).

In that role, I was constantly learning from my colleagues and the daily interactions with medical professionals and scientists from various disciplines. It was an incredible experience. We proudly worked together as one united team, impacting public health, a manifestation of my dream! What I enjoyed most was the collaborative work environment. I always came home inspired and ready for the next day.

After a few years in this role, I was selected for the Leadership Readiness Program, where only 20 employees among 1,800 were trained as leaders. It helped me realize my desire to expand my influence on public health. I wanted to help patients with diseases with limited treatment options (unmet medical needs). I knew I needed a role in public health policy to reach this goal. My tenacity landed me an offer from the Office of In-Vitro Diagnostics and Radiological Health (OIR) at the FDA, allowing me to help shape how or if new devices come to market.

I felt a new love for my work now that I could tie together my training, skills, and passion for advancing public health. Speaking with patients and healthcare professionals, like doctors and nurses, who used these devices provided vital insights into their needs. These insights guided my work to ensure patient needs were met.

After 12 years at the FDA, I can tell you that the medical device industry is essential in advancing public health. My next job was at a medical device company as the director of regulatory policy. The position allowed me to shape and influence regulations that directly impact public health. I was on the leading edge of healthcare innovations, providing the US with new and reliable medical devices and solutions. Another dream come true! After three years in that position, I was recruited by a major medical device company to serve as the director of global regulatory policy, giving me the chance to extend my impact beyond US borders. I'm grateful to have a career that is rewarding and fulfilling. All my life and work experiences have prepared me to realize my dream of contributing to public health. I strive to positively influence society with my attitude, passion, and tireless dedication to my work.

"Never let anyone tell you that you are not worthy of learning and achieving greatness."

F. RAZJOUYAN

The challenges I faced throughout my journey were significant, but as my mother always says, no one in this world is without sadness and challenges, some bigger and some smaller. What matters is your ability to adapt, persist, and never give up on your passions. Never let anyone tell you that you are not worthy of learning and achieving greatness. Always work hard towards your goals and apply discipline and consistency. When you get there, reach back to hold someone else up because by lifting others, we also rise; cherish the joy and satisfaction that follows. During my most crucible moments, I remember that everything happens by the grace of God; everything that I have is by the grace of God. Then, I visualize myself as water and the challenges and roadblocks in front of me as rocks. Water always finds its way, passes through rocks, even the very big ones, and comes out purified. This visualization helps me, so I want to share it with you. Be like water, continually moving towards your true purpose and coming out of challenging situations stronger than before. It helps to be surrounded by people who empower and support you no matter what, people who appreciate you for who you are. I want to end with a quote I hold dear: "Find the purpose; the means will follow."

"Always work hard towards your goals and apply discipline and consistency to reach them. When you get there, reach back to hold someone else up because by lifting others, we also rise; cherish the joy and satisfaction that follows."

ACKNOWLEDGMENTS

First and foremost, this book would not have been possible without the amazing women who shared my vision for this book, graciously contributed their stories, and entrusted me with them. Your passion, tolerance, and praise were genuinely humbling.

To my children, Gala and Hans Peter, the inspirations for this book, thank you for sacrificing our family time and inspiring me to make the world a better place for you. Gala, thank you for always believing that I can do anything I set my mind to.

Thanks to my husband, Todd, for providing me with a supportive environment that enabled me to complete this work, tolerate my long hours, and allow me to live like a hermit on the weekends for those long years. By the way, we can officially tell the neighbors that I am not in the witness protection program. Without you and my mom, Rose, we would have gone without dinner many nights.

I want to thank Cindy Simpson, the Association for Women in Science, Jennifer Scott, and the Society of Women Engineers for giving me a platform to find contributors. Even though we were strangers, you believed in me and my dream. Your early support fed my enthusiasm and helped make this book possible.

A very special thanks to Kathy Bowen, Auburn Cole, Dean Eklof, and their students for allowing me to test drive my stories. I cannot thank you enough for the feedback I received; I am indebted to you. Anne Camille Talley, it was your idea in the first place. I am very grateful for your advice and encouragement.

Thank you to Daria Mark, Brian, and Daniela Bigda for your technology skills, advice, and encouragement.

To my work family, LeeAnn Ali, Jayne Macedo, Craig Wisman, and Amanda Resendes, for being my cheerleaders over the years, no matter what crazy ideas came into my head. Don't worry, I have more!

My gratitude to Ashleigh Kyle for her expert legal reviews and practical viewpoints. You are an inspiration.

To delve deeper into the journeys of the phenomenal women who contributed their stories to this collection and discover additional resources, scan the QR code on the back of the book or visit dawnheimer.com. Connect with us online and join us in empowering young girls to become tomorrow's STEM leaders!

YOUR THOUGHTS MATTER TO US!

Thank you for joining us on this journey!

We hope you enjoyed Determined to be Extraordinary and found the adventures within these pages inspiring. Your feedback not only helps us improve, but it also assists fellow readers in discovering new reads that resonate with them.

Would you share your experience?

Taking a moment to leave a review means the world to us and the community. Whether it's a line or a paragraph, your insights are invaluable. Plus, it's super quick and easy!

Here's how:

1. Simply **scan the QR code below** with your smartphone camera or QR code scanner app.
2. This will take you directly to our review page.
3. Share your thoughts about the book, what you loved, and what stayed with you.

Alternatively, you can visit our website:

DawnHeimer.com

Reviews not only brighten our day, but they also enable us to keep creating content that you'll love.

Thank you for taking the time to leave your mark on our story.

Warmest regards,

Dawn Heimer
Editor